I0147899

The Mysterious World of the

BULL KELP FOREST

The Mysterious World of the BULL KELP FOREST

JOSIE ISELIN

ILLUSTRATED BY ELLEN LITWILLER

Heyday, Berkeley, California

Library of Congress Cataloging-in-Publication Data is available.

Cover Art: Ellen Litwiller
Cover Design: Archie Ferguson
Interior Design/Typesetting: Josie Iselin and Ashley Ingram

Published by Heyday
P.O. Box 9145, Berkeley, California 94709
(510) 549-3564
heydaybooks.com

Printed in East Peoria, Illinois, by Versa Press, Inc.

10 9 8 7 6 5 4 3 2 1

To the growing cadre of kelp forest enthusiasts—scientists, artists, and all those walking the shoreline, exploring the tide pools, and falling in love with all that is beyond and below.

ABOVE
BELOW

In 2021, Marianna Leuschel and Josie Iselin launched an ocean literacy campaign, Above/Below, and in late 2023 brought to life the web-based book titled *The Mysterious World of Bull Kelp*. The highly visual and rigorously designed website has been the pillar around which the Above/Below campaign has swirled, including a series of Kelp! art & science exhibits and an annual KelpFest! bringing the bull kelp story to the communities of California's North Coast through art, science, food, and film. This book is an extension of the webstory. They go hand in hand.

Please visit *The Mysterious World of Bull Kelp*
on the web at bullkelp.info

Contents

Preface

When ocean temperatures are cold, walking the beach in Northern California, Oregon, coastal Washington, British Columbia, or Alaska means encountering great piles of twisted bull kelp along the shore, especially after winter storms. Or it might be a lone mature bull kelp thrust upon the sand, its long single stipe a uniformly thin thread of kelpy material, paced off at 30–40 feet from holdfast to bulb. Where this remarkable amount of biomass comes from is usually a question mark; the actual kelp bed is hidden offshore, out of sight. Divers and snorkelers get to explore these underwater forests, swimming amid the golden blades and snaking bull kelp stipes, feeling the smooth kelp tissue in heavily gloved hands, and watching sunlight filter through the fronds while encountering fish, seals, and plentiful invertebrates, such as crabs and anemones, on the sea floor. But those of us who do not dive need to learn in different ways.

I am a beach walker, and the enormous piles of twisted kelp wrack encountered along the sandy stretches of Fort Funston in my home city of San Francisco, or just up the coast on Limantour Beach at Point Reyes National Seashore, led me deep into the

world of bull kelp. I placed bull kelp, both huge and tiny, on my scanner, making portraits of the individual specimens. I learned about its complex life cycle and amazing physiology from the seaweed taxonomist Kathy Ann Miller. *Nereocystis luetkeana* is the second chapter of my book *The Curious World of Seaweed*, released in 2019. I wrote about the history of its science and how the fate of sea otters affected the fate of the kelp forest. I learned how tribes used bull kelp stipes as fishing lines, and found the papers of Edna Fisher, who observed the sea otters off Big Sur in the 1940s and wrote seminal papers about their behavior in the bull kelp forests there. As I was writing that chapter, the decline of bull kelp along the north coast of California was starting to make the news, and I knew I had to dive deeper into the world of bull kelp. I was working on this book proposal when I met Marianna Leuschel. Together we launched Above/Below, an ocean literacy campaign, and brought *The Mysterious World of Bull Kelp* to life as a web-based book at www.bullkelp.info in late 2023.

The content of the written stories comes from a decade of connecting with bull kelp experts from up and down the West Coast. My original intent was to snorkel in bull kelp beds in each of the eight regions described, but COVID happened. Instead, we worked with intrepid and committed underwater photographers who bring the bull kelp's story to life in their challenging medium. The network of people involved with bull kelp has grown over the past five years, and it is a highly collaborative and supportive community. As a group, we share a love for this particular kelp, with its long singular stipe, tough round bulb, and golden blades—a remarkable kelp that is an essential element of this stretch of the northern Pacific Ocean's edge.

This book is a new version of the webstory; they are companions. The website has many linked resources as well as a section explaining kelp restoration methods and case studies. You can dip into it from different starting points, which provide diverse vectors of understanding. This book, on the other hand, is an intimate story to be enjoyed without a screen, on the beach, curled up in a cozy chair, or in a library. It will introduce you to a cast of characters recognized as key players in the great drama that unfolds beneath the waves, starting with the protagonist, the bull kelp, followed by its antagonist, the sea urchin, and then the sea urchin's predators, the sea otter and the sunflower sea star. Rockfish, a commonly fished-for delicacy, is an important kelp forest resident to know because of its remarkable long life. We include a section on kelp wrack—what we find washed ashore—as well as sections on shorebirds and salmon, those fish that connect the forests on land with the forests in the sea. We humans come at the end of the list, not because we are different than the other characters in the kelp forest drama—we have been in it since the beginning—but because this placement suggests where we are headed, and how we humans can rethink our relationships to ocean organisms. The beauty of the kelp forests can spark our curiosity and encourage us to venture into unfamiliar worlds, out onto the beach, and beyond.

The Mysterious World of the
BULL KELP FOREST

Introduction

There is a sliver of the oceans adjacent to the continents known as the "sunlight zone." These are waters deeper than we can explore at low tide but where sunlight still penetrates down to the ocean floor. The cold, nutrient-dense waters of the North Pacific sunlight zone that follows the coast from Central California through the Pacific Northwest into Alaska enable ocean forests to flourish, hidden offshore out of sight. These are the kelp forests. Like forests on land, the kelp forest is inhabited by a plethora of organisms, and in this region of the Pacific, the dominant kelp is *Nereocystis leutkeana*, or bull kelp.

Bull kelp is a magnificent organism. It can grow 10 inches a day, from a tiny sapling as delicate as a piece of golden tissue paper to a hefty 35 pounds of pure algal biomass rising 60 feet from the ocean floor to surface in just a matter of months—the redwoods of the North Pacific sunlight zone. It is surprising, opportunistic, undervalued, foundational, resilient, and vulnerable. It is the kelp within which herring school and spawn, rockfish abound, and seals seek prey. It is the canopy under which countless other algae grow, including the understory kelps and pink rock-encrusting corallines that attract the settling of urchin and abalone. Under the gestural swaying largess of the bull kelp is a colorful world with starfish, nudibranchs, and sponges. There are other large kelps in the mix—giant kelp, feather boa kelp, and bladder-chain kelp—but from Central California to Alaska, bull kelp is the major player, massive and hefty, thriving in the nutrient-rich, cold waters of the northern Pacific.

There are grand and majestic kelp forests along the temperate edges of all the world's oceans. Giant kelp dominates Southern California, down the Baja Peninsula and around both coasts of South America. South Africa has famous kelp forests, as do South Australia, New Zealand, and Scotland, all with their own dominant kelp species and suite of associated seaweeds. These ocean forests are as diverse and abundant and important for life on earth as any of our terrestrial forests. All kelps grow just a few hundred yards from shore in what is known as the subtidal zone, needing a rocky bottom to hold on to and enough sunlight to promote growth. Kelps are profoundly efficient at uptaking carbon dioxide and water and, with the spark of sunlight and the nutrients of a cold ocean, growing at a remarkable speed. But of all the kelps, bull kelp is perhaps the fastest grower of them all, reaching maturity in a matter of months from late winter through summer. This growth is primary production, the underpinning of the kelp forest food web.

The continual sloughing away of the kelp blades creates a snowstorm of organic matter that feeds countless organisms, such as shrimp, isopods, snails, crabs, sea urchins, and abalone. These low-level detritivores are important food sources for herring, rockfish, juvenile salmon, and lingcod, and then these in turn build the food web important for marine mammals such as seals, sea otters, and whales as well as seabirds diving into the kelp forest from above. Kelp detritus thrown onto the beach, known as wrack, is decomposed by kelp flies and amphipods, which are an important protein source for shorebirds. The entanglement of interdependencies is known as the trophic food web—who eats whom—and bull kelp is the foundation of this web in the coastal waters of the northeastern Pacific Ocean.

The relationships among these varied organisms within the bull kelp forest, whether scaly fish, spiny urchin, or flexible kelp, create an ever-shifting drama that plays out in unique ways in different regions along the bull kelp's range. While bull kelp is a coast-wide species, for us to understand the regional kelp forest dramas (and especially the ones nearest home, the ones we care about) it is important to be familiar with the cast of characters. They are important to know about in their own right, but also because they illustrate some of the connectivity of the kelp forest food web. By diving deeper into the life history and historical ecology of a few of these primary actors of the bull kelp forest, we can learn how humans, both ancient and contemporary, have affected the balance of the kelp forest over time, and learn about ocean systems and how changes are affecting these very different kinds of beings. The drama in the kelp forest is always fascinating.

We *Homo sapiens* have been players in the kelp forest drama since our species first wandered the coastlines following the "kelp highway," benefiting from the mineral- and protein-rich abundance of shellfish and seaweed. We humans have been and still are stewards of the coastal bounty, but also disruptors of balance and destroyers of resilience. We have also been teasing out the relationships among the kelp forest players—we call this ecology—for decades. But this attention to understanding kelp forests represents a mere blip in time compared to the hundreds

of years of learning about terrestrial plant ecology and thinking about trees and forest health. The newness of the exploration as well as the remoteness of this underwater world, especially that of the bull kelp, keeps mystery alive and well as a formidable player in our human understanding of the kelp forest.

Because these offshore glades are hidden underwater, many people only hear about a kelp forest along their particular stretch of coastline with news of its decline or disappearance. We emit a collective gasp at our feeble comprehension of these complex and essential worlds only as they fade away. There are many committed scientists, tribal ecologists, marine resources committees, students, divers, volunteers, and more working feverishly to understand the inscrutable mechanisms of bull kelp growth and kelp forest dynamics in order to lend a human hand to rebuilding resilience and helping regain the biodiversity of a healthy kelp forest. But those people focusing on bull kelp are at the beginning of kelp recovery work. There are places around the globe that have been at it longer than we have. Korea has a profound cultural connection to kelp, and the state has devoted years, space, and money to kelp restoration in their coastal waters, especially for *Undaria*, the main ingredient of the soup Korean mothers drink upon the birth of a child and thereafter on birthdays, for health and well-being. In southern Australia, Indigenous communities have been models for leadership in restoring swaths of the historic kelp forests around Tasmania.

As we confront loss, and gain understanding of how to facilitate recovery, it is important to reflect on how we arrived at this point. Natural systems, including the bull kelp forest, evolved to endure crises such as volcanic eruptions, earthquakes lifting the sea floor by twenty-five feet, and two-hundred-year storms that dislodge kelp beds from their rocky holds, as well as fluctuations in ocean temperature. Such mechanisms as radical biodiversity, the redundancy of top predators, cold nutrient-rich ocean water, and long-term ocean dynamics have enabled the bull kelp forests along the Pacific coast's edge to rebound and rejuvenate following catastrophic events. That is what resilience is. Indigenous tribes and peoples worldwide have lived alongside kelp forests and harvested from the nearshore ocean waters for millennia, sustaining resilience through their ecological practices and deep relationships of reciprocity with the plants and animals of both land and sea.

But systems of commerce and colonialism in just the last few hundred years have overtaken these relationships. The Western notion that humans are somehow outside of nature and have the prerogative to exploit the abundance of natural ecologies has become the accepted practice. However, when stressor after stressor is placed on these oceanic systems, when the concept of "sustainable" implies just enough rather than more than enough, and with the ultimate environmental stress of climate change, the complex networks of interdependent life have an increasingly difficult time rebounding. The alarming decline of bull kelp forests in some areas is evidence of this. Everywhere around the globe, kelp forests are rapidly diminishing; they are highly susceptible to rising water temperatures and other stressors we've placed on the oceans. In many locations, overpopulations of sea urchins graze the kelp down to a spiky wasteland known as an urchin barren. Sea urchins are a natural part of every kelp forest, but when the balance is lost, kelp can disappear. A new worry is that our current generation of ocean lovers hasn't experienced the natural biodiversity that rich kelp forests provide, so an impoverished coastline becomes the new normal. The complex web of interrelationships that is the healthy

kelp forest not only generates an abundance of fish and other organisms that we humans depend on for food and forage; it also creates carbon cycles that mitigate ocean acidification, boost the health of nearshore ecologies, keep ocean cycles balanced, and moderate sea-level rise.

So if we consider bull kelp as kin and understand how foundational it is to so much we hold dear, restoring the resilience of bull kelp forests can be a model for work elsewhere, and a network of these efforts can help reverse this trajectory of mutual decline. Bull kelp restoration is in an experimental phase, but there are examples of kelp restoration around the globe with a growing community of dedicated kelp lovers committed to it. The international Kelp Forest Alliance is connecting kelp restoration communities around the world so that best practices, victories, and failures can be learned from by all. The efforts along the eastern edge of the Pacific are part of this network of learning. We hope that this particular organism, bull kelp, once understood and appreciated, revered for its beauty and its resourcefulness, can be an inspiration to us humans, and serve as a guide for how to create foundations of support for the world around us. The bull kelp can be a talisman for rewilding our human mind.

Bull Kelp Community

The trophic food web embodies the complexity of who eats whom in the bull kelp forest. At the bottom are the photosynthesizers: the kelps, seaweeds, and microalgae or diatoms. They are considered primary producers and are crucial to the health of the web above them. Sea urchins and abalone eat algae, as do thousands of small invertebrates like snails, limpets, and isopods, which are in turn eaten by sea otters and sunflower sea stars. Fish residing within the kelp forest are prey for larger mammals, such as seals. The complexity of the web implies balance and stability of the kelp forest regime.

A trophic cascade occurs when the effect of displacing one of the trophic (food) levels ripples throughout the ecosystem. Sea otters, sunflower sea stars, sea urchins, and kelp form a classic trophic cascade; urchins eat kelp, and sea otters and sunflower sea stars eat urchins. The removal of both top predators (sea otters and sunflower sea stars) most often causes a systemwide disruption because of the resulting increase in urchin densities. These urchins then overgraze the kelp until the ocean floor is

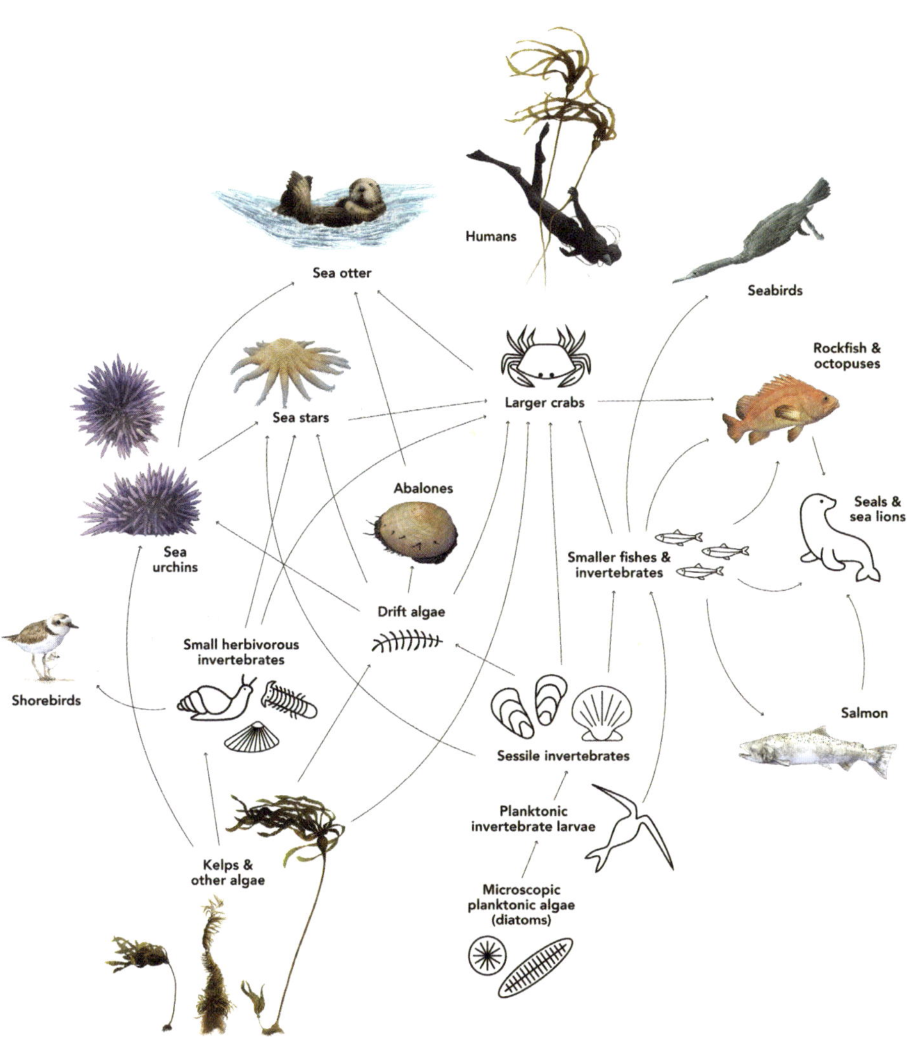
Humans
Sea otter
Seabirds
Rockfish & octopuses
Larger crabs
Sea stars
Abalones
Seals & sea lions
Sea urchins
Smaller fishes & invertebrates
Drift algae
Small herbivorous invertebrates
Shorebirds
Salmon
Sessile invertebrates
Planktonic invertebrate larvae
Kelps & other algae
Microscopic planktonic algae (diatoms)

Sea Otter
Sea stars
Sea
urchins
Kelps &
other algae

covered in spiky urchins with no kelp. This disruption in the balanced food web can cause a regime shift from kelp forest to urchin barren. Everyone is important!

The bull kelp forests of the North Pacific are a fantastic lens through which to observe these complex interactions among predator, prey, and primary producers. The trophic food web's complexity affords it strength, but when links are broken and the web is simplified, the effects on an ecosystem can be profound and hard to reverse. The kelp/sea urchin/sea otter and sunflower sea star trophic cascade has played out over a century and a half on the California coast in intriguing ways—a story that includes humans in the top predator position as well.

The sea otter's disappearance due to the fur trade allows an interesting look into human relationships to the riches and bounty (aka biodiversity) of the kelp forest along the northern Pacific coast. Without otters, the red abalone of California's north coast were predator-free, eating the plentiful and nutritious bull kelp and other drift algae. Their numbers and size increased. Starting with the "abalone rush" of the 1850s and continuing into the twentieth century, California's coastal communities (of humans) became preoccupied with foraging for abalone in the intertidal zone and, just like sea otters, free diving for them. By the 1990s, sea otters had been missing from the coastal landscape in so many places for so long that their role in the narrative had been forgotten by most humans. The intertidal ecology without them, replete in clams, abalone, urchins, and crab, came to be understood as a new standard, a shifted baseline. Since the early 1900s, up and down the West Coast, fisheries, economies, lifestyles, and expectations of coastal communities of all sorts arose from the abundance of shellfish in the absence of sea otters. By the 2010s, humans as top predator had done poorly by the abalone of the California coast. With overfishing and the disappearing food source (kelp), red abalone numbers along the Northern California coast plummeted. Recreational abalone divers became concerned for the rich, colorful underwater world they saw disappearing. The power of the bull kelp forest to sustain the things they loved was clear—once it was gone.

Unlike abalone, the purple sea urchin of the bull kelp forest had a remaining predator, the sunflower sea star, keeping the urchins' numbers in check, despite the loss of sea otters. The larger and abundant red urchin, however, became a sought-after shellfish for human divers in the late 1980s and 1990s. The introduction of sushi into American cuisine and the booming Japanese economy created a strong market for urchin roe, and a red urchin fishery in California and Oregon was initiated. Humans weren't interested in the smaller purple urchin, but urchin divers became the top predator for red urchins. Urchin divers roamed the ocean floor and hauled thousands of red urchins to the surface, into the harbors where they were processed and shipped to market. These urchin divers reveled in the glory of being in the bull kelp forest. During those years when divers kept red urchin numbers down and sunflower sea stars kept purple urchins scared and staying in their hiding places, the bull kelp, despite its naturally cyclical abundance, enjoyed a bonanza period. In 2008 there was so much bull kelp in Van Damme State Park on the Mendocino Coast, a bird could walk from the south shore of the bay to the north shore without getting its feet wet, picking its way easily on the tubular bull kelp lying packed together on the water's surface. Urchin divers and abalone divers waxed rhapsodic about the beauty and majesty encountered underwater in the kelp forest. Then they were the first to see it change.

In the first months of 2014, anyone diving at the ocean floor in California and the Pacific Northwest noticed the sunflower sea stars melting away. A devastating wasting syndrome hit all of the sea stars up and down the Pacific coast, but it hit the *Pycnopodia* the hardest and with the most prolonged effect. Billions of the large, twenty-armed urchin hunters disappeared. Around the same time, two ocean warming events occurred. Warm water holds fewer nutrients than cold water and puts huge stress on kelps and seaweeds. Bull kelp's vast nutrient needs could not be met by the impoverished ocean. The reduction in kelp, together with the lack of predatory pressure, prompted the purple urchins to emerge from their cracks and go searching for food. They munched down whatever kelp and seaweed they could find, clearing rocks of all life except the purple spiny echinoderms. It happened quickly—out of sight to most people.

Ecologists, however, are using the wide-ranging kelp forest food web links to track the kelp forest from onshore. Seabird nesting colonies are relatively easy to document, and the breeding success of seabirds is a good indicator of general ocean health. If the small fish these seabirds prey on are plentiful in the waters adjacent to the rocks and cliffs where they nest, baby birds fledge, grow to adulthood, migrate by sea, and return to nest, lay eggs, and complete the cycle. But it all depends on the kelp forests' largesse as fish habitat. Work overlaying long-term satellite data imaging of kelp abundance with long-term seabird population data sets is making this kelp–seabird intimacy explicit.

BULL KELP

Nereocystis luetkeana

Nereocystis luetkeana is the only species in its genus, but it is in the Laminaria family with other kelps such as giant kelp, sugar kelp, and sea palm. Seaweeds, or marine algae, can be either green, red, or brown. *Kelp* is not really a scientific term, but everyone uses it for big, fleshy marine algae that tend to be shades of golden or olive brown. Kelps are a subset of the class Phaeophyceae (from the Greek, meaning "dusky" or "shadowy")—the browns. No other seaweeds but the kelps have gas-filled bladders, or pneumatocysts, to keep their bulky kelp bodies and blades afloat.

Range: Central California to the Aleutian Islands

Kelp are the largest seaweeds with the most differentiated bodies, and bull kelp has, perhaps, the most extraordinary form of all the kelps. The bull kelp's single stipe is a long, uniformly thin rope of tough cortical material that can stretch up to 38 percent to give and take with the ocean current. It emerges from a small holdfast anchored to a rocky substrate and stretches to a single bulb, or bladder, shaped like an oversized avocado. The blades, streaming out from the bladder—or pneumatocyst—multiply at four source points into as many as sixty ribbons of thin golden fronds. In springtime, baby kelp in various stages of development reach for the surface, their golden bladders no bigger than your thumbnail, their four golden blades catching the sunlight, radiating its wonder in their molten translucence. These young *Nereocystis* use the power of sunlight and the nutrients of the ocean to perform one of the greatest feats of metabolic growth on our planet: They will become a massive and majestic bull kelp in only a matter of months.

Nereocystis luetkeana is, generally speaking, an annual. The entire organism grows anew each season, typically starting in late winter or early spring, and for six months that growth is astounding. The spring juvenile stipes are thin and uniform, but as the kelp matures, the long singular stipe (or stalk) grows in a reverse taper, from thin at the bottom to thick toward the top. It can grow 6–10 inches a day and reach heights of 60 feet in a matter of months, slowing once the bladder senses proximity to the surface. This stupendous growth is all in the service of efficient photosynthesis—getting the blades closer to sunlight to catch as many photons as possible. At low tides, the upper, tubular portion of the mature bull kelp stretches out along the ocean surface like a passive snake undulating with the swells, its blades hanging beneath. After about two hundred days, or with the onset of winter storms, the entire organism dislodges from the ocean floor, holdfast and all, and washes away, sinking as it breaks up or tumbling onto sandy beaches. Bull kelp grows in clumps, often with stipes entangled, so when one bull kelp's holdfast is ripped from the ocean floor, the entire group breaks away and washes ashore as a knotted mass. The flotsam can be voluminous snarls of hose-like, dark tubes with bladders jutting out in all directions. This bull kelp wrack in turn fuels other food webs. Some individual bull kelp, produced late in the season, may successfully overwinter and survive a second year, up to eighteen months. These are old-growth bull kelp grandmothers, usually with plenty of other algae (epiphytes) and bryozoans (a lacy crust of colonizing animals) growing on them.

Bull kelp's spore patches are the acorns of the kelp forest. The extended daylight of summer, combined with the upwelling of nutrients, accelerates bull kelp growth, and by the end of summer, a dark chocolate-brown patch 2–7 inches long emerges on each blade. Called sori (plural), these fertile patches are bundles of millions of spores. As growth continues, the maturing sori gravitate to the outer end of the blades, while new ones emerge closer to the bladder. Coordinated with the light of dawn, a whole sorus (singular) will detach from its blade, falling away to float to the ocean bottom and settle near the base of its parent. Within hours, the sorus releases millions of photosynthetic spores into the water column to disperse, the majority settling on the rocky bottom. With forty to sixty blades each, an adult *Nereocystis* can produce over three trillion spores. By the time fall arrives, only the ragged edges of the blades remain, hanging like leftover dough after cookies have been cut out. Soon the entire adult kelp will loosen from the bottom and wash ashore.

The tiny spores that successfully settle on the ocean floor

germinate into microscopic, thread-like sexual organisms, males and females, also known as gametophytes, which sport only a single set of chromosomes. These tiny organisms lodge in the cracks and crevices of the ocean bottom to wait out the rough weather and dim light of winter.

While cultivating bull kelp in the lab is not difficult, studying this cryptic sexual phase in its natural environment is virtually impossible, and questions abound. It is not clear how long these filamentous microorganisms can persist. We do know that when the time comes, the sessile (fixed in place) female releases a powerful pheromone that attracts the nearby roving sperm, and fertilization initiates the development of a baby sporophyte that grows into the massive bull kelp we recognize. This new growth begins in late winter or early spring, and the annual cycle begins again. The rotation of kelp on the beaches each year indicates the rhythms of robust growth out under the waves. Piles of bull kelp on the beach in late summer to midwinter signify healthy kelp beds offshore. This is part of the bull kelp single-year life cycle.

Most beachcombers encounter bull kelp as wrack, strewn on the beaches in enormous tangles of knotted stipes and toughened bladders. Sometimes other kelps—feather boa or giant kelp—are caught in the morass, but often enough it is pure bull kelp tangle. When the holdfast comes along and is still attached to the twisted stipe, we can see that it is a modest anchor for so much biomass. Storms rip bundles of mature bull kelp from their watery home toward the end of a healthy growing season, leaving entire clumps at the mercy of the waves. Another form of bull kelp wrack found on spring beaches is the tiny baby bull kelp glinting at the water's edge. These are magical and precious signifiers of the new kelp growth in some unseen bed out beyond the waves.

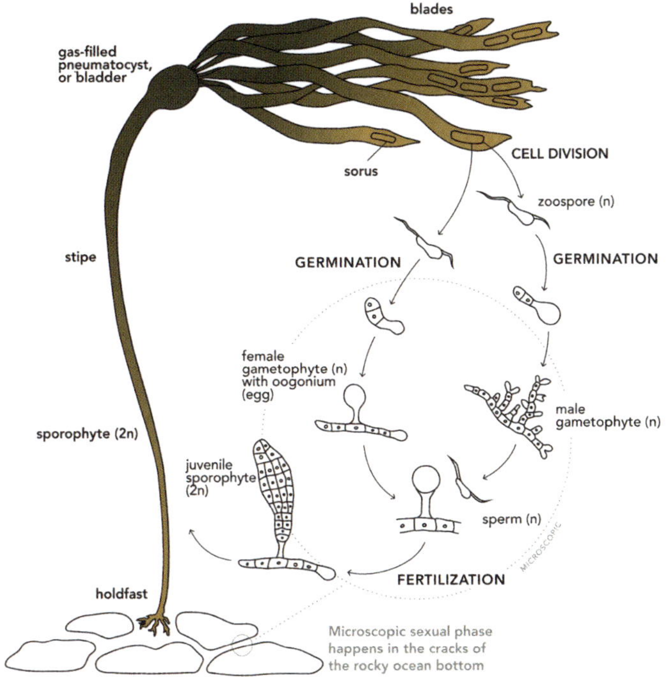

Diagram after Mondragon showing the alternating sexual (microscopic) and asexual (sporophyte) generations of the bull kelp life cycle. Bull kelp is an annual and starts this cycle afresh every spring.

The wrack line, where the high tide has left its debris, is a zone of treasure to be explored and investigated. Kelp wrack can tell many stories. Nori, generally, is the red algae that usually grows on rocks in the intertidal zone, delicious to snack on, but a specific nori epiphyte grows only on bull kelp and can often be seen as a red beard growing on an old mama bull kelp that has washed ashore, suggesting that this kelp lived well past a full year. Wrack distribution in some coves and not others reveals clues about currents and tides. Often the source of bull kelp wrack on a given beach is unknown, with no apparent kelp beds visible offshore. In fact, there are genetic markers, known as environmental DNA (eDNA), that can identify kelp by region, and scientists are using these to study the genetic diversity of bull kelp populations up and down the coastline. As bull kelp abundance ebbs and flows, researchers are mapping its genetics to gauge trait variation and health.

Bull kelp wrack on the beach, in various stages of rapid decay, reminds us that kelp is not necessarily an effective carbon sink. Kelp, and especially bull kelp, are remarkably efficient at pulling CO_2 out of ocean water and transforming it into biomass. This is the power of photosynthesis. But unlike the trees of a forest on land, kelp is constantly desiccating and degrading, shedding and sloughing. Bull kelp's biomass is only in existence for a short while before it and its carbon degrade back into the ocean waters to become part of the complex food web around it. Kelp on the beaches breaks down fast as well, consumed by insects to become part of the nearshore ecologies. Bull kelp's state of constant transformation is perhaps not helpful when thinking about fixed carbon sequestration, but does serve as a reminder of how dynamic the kelp forest is in cycling nutrients and carbon through food webs and shoreline ecosystems—extraordinarily important for the health of a panoply of species and their interrelationships.

Great expanses of sandy beaches punctuate the edge of the continent adjacent to the bull kelp forests of the North Pacific, between rugged outcrops, rocky cliffs, and bluffs. Bull kelp and other kelps wash up onto these beaches, creating piles of wrack that perhaps aren't recognized for the enormous benefits they provide to the entire nearshore food webs. They are known as trophic subsidies to the beach and dunes, a nutritional bridge between the marine world and the world onshore. In their afterlife, bull kelp and other kelps form the base of the food chain

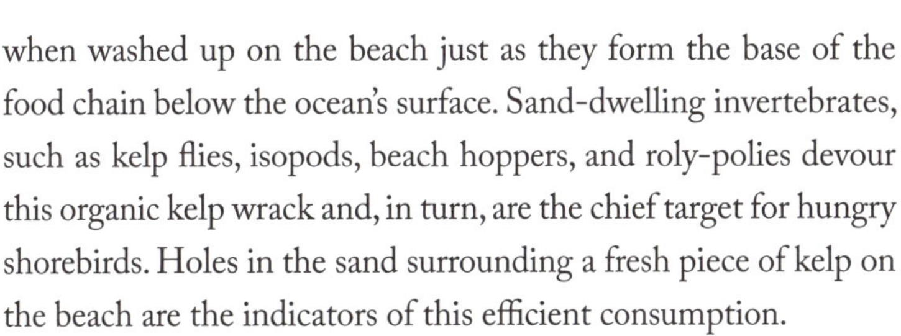

when washed up on the beach just as they form the base of the food chain below the ocean's surface. Sand-dwelling invertebrates, such as kelp flies, isopods, beach hoppers, and roly-polies devour this organic kelp wrack and, in turn, are the chief target for hungry shorebirds. Holes in the sand surrounding a fresh piece of kelp on the beach are the indicators of this efficient consumption.

Long-term studies have shown a one-to-one relationship between the amount of kelp wrack on the beach and healthy shorebird populations; kelp detritus means (tiny) invertebrates, which in turn mean healthy birds. How shorebirds find these juicy crustaceans and kelp flies depends on what kind of predator they are. Some, like the snowy plover with its big eyes, are visual predators; other shorebirds are tactile foragers, probing and sensing with their long, sometimes curving bills. In either case, the bull kelp's connections from sea to shore are clear. The shorebirds flitting among the wrack are showing us.

PURPLE SEA URCHIN

Strongylocentrotus purpuratus

Strongylocentrotus purpuratus is one of three sea urchin species that inhabit the nearshore waters of the Pacific. The green sea urchin is in the same genus. It is found farther north from Vancouver Island through Alaska. Red sea urchins, larger than the other two, are in another genus; they are found throughout the bull kelp's range and are fished for their roe, or uni.

Range: Baja California to British Columbia, Canada

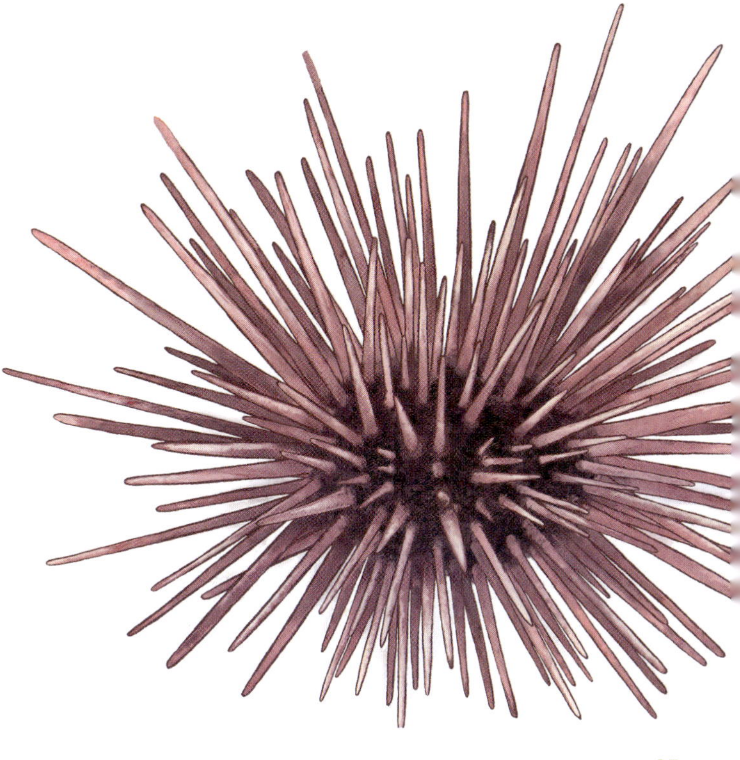

Sea urchins are part of all the world's kelp forests. They can be found in ocean waters from the edge of the low-tide waterline through the subtidal kelp forest zone and even deeper still. They are known as echinoderms, a phylum shared by sand dollars, sea stars, and thousands of other species. If you compare all three—urchin, sand dollar, and sea star—you will notice their shared five-fold radial symmetry. The sea urchin's beautiful domed structure is made of calcium carbonate, but it is not a shell; it is known as the test. A sea urchin's mouth has five radial, bony teeth in its underside that can scrape and nibble—and what it likes to nibble is kelp and seaweed. Sea urchins are herbivores, like cows, but they graze on the primary producers of the oceans: the algae—that is, kelp and seaweed. Urchins are voracious. They eat a lot.

Bull kelp and its understory seaweeds are constantly sloughing off at their margins, the currents and constant forces of the ocean waters pulling the degenerating tissue from the edges of the kelp even as new healthy tissue is created. This algal detritus floats in the water column, making it murky and hard to see through—"low vis," as divers say. But these murky waters are indicators of a healthy ocean, rich in food material floating along for urchins and other detritivores to profit from. Urchins typically live healthy existences lodged in one place, often a niche or alcove created in the rock, eating the detritus that floats by. When a kelp forest is in balance, with ample algal material floating around, urchins have plentiful innards, or gonads (what we know as urchin roe or uni, the golden stuff eaten as sushi), which provide a good source of protein for their predators, the sea otter and the *Pycnopodia* sunflower sea star. These predators keep bull kelp forests in balance. If systems shift and there is not enough floating algal material, the urchins can get up and move on their suctioned tube feet that emerge between their spines and go on the march looking for healthy kelp. This can happen when there is less kelp and seaweed around, whether because of warming oceans or because of an increased number of urchins relative to kelp abundance, as happened when the sunflower sea star was eradicated by disease.

When urchin populations are not kept in check by predators, kelp forests can transform into urchin barrens—bare rock bristling with carpets of purple and red urchins with no kelp in sight. This regime shift, as it is called, can be swift. Regions of lush historic

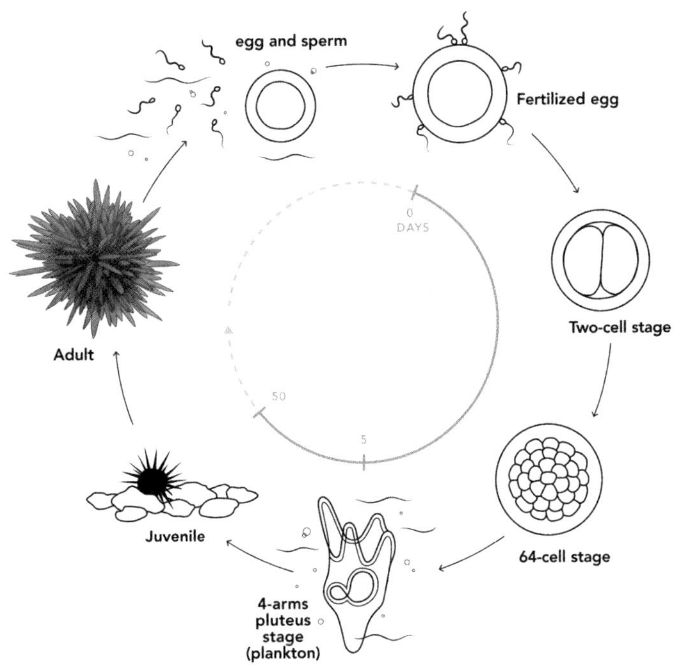

Sea urchin life cycle: Sea urchins are broadcast spawners. Each urchin ejects millions of eggs or billions of sperm simultaneously (it is impossible to tell male from female) into the water column through holes in the top of its domed test. The chance fertilization of eggs by sperm results in planktonic larvae that eventually settle and grow into long-lived urchins.

bull kelp beds have been decimated by these hordes of urchins. Young bull kelp reaching for the surface along the Northern California coast in spring are eaten by the purple urchin before they can reach maturity and reproduce. But once they've eaten the algae, urchins do not die of starvation and open up space for new kelp to grow; they hunker down, lower their metabolism, and wait. They use all the reserves stored in their gonads and continue to exist as marginal creatures, empty, domed, and spiky husks that are, nevertheless, not dead. Zombies perhaps.

Humans are another key predator of the sea urchin. Red urchin grow old and large—up to two hundred years old—and the uni they produce hold great value to fishers. Their populations are kept in check by the market. The smaller purple urchin, however, has never had market value. Their lack of value as a commercial fishery, and in the absence of other key predators in the bull kelp forests of California's north coast and other regions, has allowed their populations to explode. There are myriad efforts under way to create a market for these urchin, whether as soil enhancements or even as ranched urchin. The zombie urchins are fattened up in tanks on land—fed kelp and seaweeds—until their uni are acceptable to high-end chefs and other culinary endpoints. The flip side of these innovations is a recognition of possible harm in creating a market or fishery dependent on an unhealthy natural system. As humans meddling in these ocean networks, we must be mindful to incentivize healthy balance and pathways to biodiversity.

SEA OTTER

Enhydra lutris

It is generally accepted that there are three subspecies of sea otter, a southern (*Enhydra lutris nereis*), northern (*Enhydra lutris kenyoni*), and Asian (*Enhydra lutris lutris*) sea otter, each inhabiting that portion of their range. The southern sea otter's limited population maintains its protection under the federal Endangered Species Act; the northern sea otter is protected under the Marine Mammal Protection Act.

Range: Central California (southern sea otter), coastal Washington through British Columbia and Alaska (northern sea otter), and around the islands north of Japan (Asian sea otter)

Sea otters are the largest member of the weasel family but the smallest marine mammal. They have whiskers, large webbed forepaws with retractable claws, and ears and noses that can close when diving. They have great senses of sight, smell, and touch; float mostly on their back on the surface using their tail as a rudder; and have paddle-shaped hind feet for swimming, not walking or running on land. When alarmed they position themselves vertically in the water with their head periscoping around to spot danger and warn their fellows. Sea otters are related to river otters, who will come down to the shore and are often mistaken for sea otters, but they are distinct species. Unlike river otters, sea otters can stay completely at sea, foraging, resting, mating, giving birth, and nurturing their young among the kelp fronds, or they can inhabit coastal estuaries, occasionally hauling out on sandbars like seals.

Sea otters dive day and night for their food and are limited in their habitat by what they can find to eat during the length of a good breath hold. For this reason, they share coastal waters with the bull kelp forest, which needs a similar proximity to shore for best access to sunlight. Sea otters can dive and forage to about 60 feet, holding their breath for about 2–4 minutes. They must eat about a quarter of their body weight each day to maintain a hypercharged metabolism, and although they eat a variety of mollusks and invertebrates, sea urchins are a favorite; they can eat 15–20 pounds of large urchins in a day. An otter dives to the ocean floor and, using its extraordinary sense of touch, quickly identifies a viable, uni-rich urchin, then carries it and a stone to the surface in a pouch of loose flesh under its forelegs. It places the stone on its belly and smashes the hard and spiny urchin to get at the nutritious food within.

Unlike other marine mammals, sea otters do not have blubber and therefore have an extraordinary need for calories to generate heat. They also sport the thickest fur of any animal on earth, allowing them to thrive in cold Pacific waters. This dark-brown lustrous pelage must be continually fluffed up so that the air bubbles trapped in between all those hairs can provide the layer of insulation the otter depends on. If a sea otter is not able to preen or its hairs get oily or dirty, it will die quickly of hypothermia. This is why, in addition to foraging for food—and the moms teaching their young how to dive and forage for food—a sea otter's major preoccupation is grooming its fur. A mother will keep her pup's fur in such a condition that the pup looks like a little puff ball and floats like a cork.

Sea otters rest together in groups called rafts. There are rafts off the coast of Washington state that number in the hundreds of otters. The entire population of southern sea otter in Monterey and Central California is descended from a single raft of about fifty otters that stuck together and survived in secret in the coves beneath the cliffs of the rugged coast of Big Sur until they were discovered in 1938 and protections began.

Each female will bear just one pup, and then nurse and care for that pup for 8–12 months. She teaches her otter baby to dive, feed, and groom, passing along her own food preferences, be it crab, clams, abalone, or urchins. She will go into estrus and get pregnant again shortly after abruptly weaning and abandoning her pup, and, gestation being six months, she'll give birth again at any time of year. The female otter not only has her own metabolic needs but is most often gestating or nursing a baby otter, so she does not travel far, staying close to a food source that can keep her and her newborns thriving.

Most often the females stick together, creating nurseries in the kelp beds. An otter mom will wrap her pup in a kelp frond or leave it to bob among the bull kelp tubes that enclose a patch

of ocean, while she dives for food she will share with her baby. In this way food preferences and diving techniques are passed from one generation to the next. While females stay close to home, young males are the pioneers who venture into new territory. But these are only the few. Most otters are tightly bonded to place. They know where they are from, and if brought to some other location, as various translocation efforts have done, the otters will most often try to make a beeline back to "home," back to where they were born.

Sea otters have historically been part of the kelp forest ecology of the North Pacific. It is estimated that prior to 1740, when the second Bering expedition set out from eastern Russia to explore the Aleutian chain of islands, find America, and inadvertently spark the colonial fur trade for otter pelts, around 300,000 sea otters populated their entire range skirting the North Pacific, from the very northern tip of Japan, around the northwestern edge of the Pacific Ocean, along the Aleutian Islands, down the coast of Alaska, the Pacific Northwest, and California to Baja California in Mexico. This is not a large number of marine mammals in total, but sea otters are individually impactful—each otter contributes significantly to the ecosystem it inhabits.

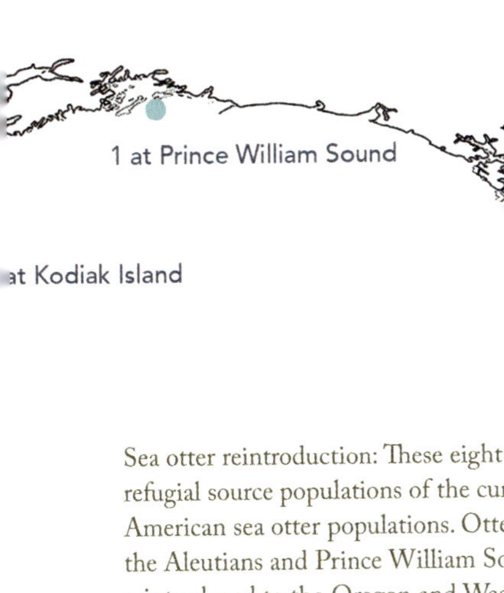

Sea otter reintroduction: These eight sites indicate refugial source populations of the current North American sea otter populations. Otters from the Aleutians and Prince William Sound were reintroduced to the Oregon and Washington coast as well as Southeast Alaska in the later 1960s and early 1970s. Oregon's otters disappeared. SE Alaska and Washington coast otters have thrived. British Columbia's otters are returning in some places.

Between 1750 and 1850, fur traders, initially from Russia and then from Spain, Britain, and America, wiped out almost every male, female, and baby sea otter there was from the Aleutian Islands south into Mexico. This violence in the oceans and on land (they also hunted fox, fur seals, and beaver) took place in concert with the settlement of the soon-to-be states of California, Oregon, Washington, and Alaska, and the province of British Columbia. The indiscriminate killings left sea otter populations functionally extinct by 1890, with only a few refugial populations (see map). One of these was at Amchitka Island, the largest of the Rat Islands, a group of windswept, treeless islands at the very lowest point of the curve of the Aleutian Islands stretching between Alaska and Russia.

Ironically, the threat of further devastation resulted in the reintroduction of sea otters back into some parts of their former range. In 1965, the US conducted the first atomic test at Amchitka Island. Another bomb was detonated in 1969, four thousand feet belowground. An even larger underground test, called Cannikin, was planned for 1971. In the face of the possible extinction of the remnant sea otter population, biologists desperately airlifted twenty or so otters to a number of mainland sites in SE Alaska, British Columbia, the outer coast of Washington state, and the Oregon coast. Oregon translocation failed; no otters survived. Southeast Alaska received otters from the Aleutians plus a few from another small otter population in Prince William Sound. It has proved to be otter nirvana, and the otter populations there have grown steadily. Washington coast's otters are thriving. In British Columbia, the translocated sea otters have slowly spread to new territory; each place they reinhabit sees a resurgence in kelp and seaweeds. As mentioned, Central California's sea otters are all descendants of the single raft of otters found off the Big Sur coast in 1938. The regional differences in sea otter populations illustrate their profound impact on the kelp forest.

Sea otters are voracious top predators; their favorite prey are sea urchins, clams, abalone, crab, other mollusks, and sea cucumbers. They are associated with the kelp beds they live within, keeping the urchin grazers in control so that the kelp beds and the richness of life they support can thrive. They are a quintessential "keystone species," affecting food webs beyond themselves. Like other top predators, otters promote resilience in the nearshore ocean systems they live within. The otter's capacity to hunt and eat urchins is a crucial factor in maintaining a healthy bull kelp forest habitat, keeping urchin populations at levels that allow delicate juvenile bull kelp to grow to maturity without being eaten. But while sea otters are agents of resilience, they are not agents of renewal. They will not forage in an urchin barren where zombie urchins, devoid of gonads, hold no caloric benefit.

Sea otters' connection to "home" along with many of their other traits make us humans feel intensely connected to them—sea otters are ridiculously adorable. Otters, with their beady eyes in a furry face with large whiskers, and often with a baby nestled cozily on their bellies, are hard not to think of as closely human. These characteristics are a way for many of us to bond emotionally to this marine mammal of our wild coastline and care about its well-being.

But the otter's role in the kelp forest ecology, its need for the same prey we humans desire, creates a fascinating human–otter dynamic that can range from true love to intense ambivalence. Market-driven fishers, Indigenous foragers, and recreational abalone divers tend to regard the otter as a destructive force in the nearshore waters. Sea otters can quickly reduce a clam population, and they remind abalone to stay in their cracks. Their impact on an ecology is immense. But as more people recognize the important role of the kelp forest relative to biodiversity, fish populations, and ocean health generally, attitudes toward the sea otter are changing among fishers. Bull kelp continues to thrive in those places where sea otter populations are healthy.

SUNFLOWER SEA STAR

Pycnopodia helianthoides

About thirty kinds of sea stars live along the Pacific coast, but *Pycnopodia helianthoides* is the largest. It is an echinoderm, related to sea urchins, sand dollars, sea cucumbers, and other sea stars.

Range: Just south of Monterey, California, through the Pacific Northwest, British Columbia, and Alaska to the middle of the Aleutian Islands. This biogeography matches that of bull kelp almost exactly.

Sunflower sea stars are the deepwater "wolves" of the sea—as big as a pizza—with 15–24 arms radiating out from a large fleshy middle. They don't have the structure of other sea stars and hang like rag mops when brought up as bycatch. They are vibrantly colorful, ranging from reddish-orange to purple, yellow, or violet-brown on the top side, and they sport more than 15,000 suctioned tube feet on their bottom side. They move swiftly (for a sea star) on these tube feet to swarm over and eat other invertebrates, both alive and dead, living in their ocean-bottom habitat. They can sense light and dark, and chemical cues guide them toward a food source. Their favorite prey is sea urchin, which can make up almost all of their diet. Below British Columbia, they eat mainly the smaller purple urchin; in BC and Alaska, they go after the similarly sized green urchin. They are an impressive predator, and, in the absence of sea otters, they can keep urchin populations in check, maintaining these thriving, biodiverse kelp forest habitats.

The bull kelp forests of the North Pacific evolved with two urchin predators maintaining ecological balance: the sea otter and the sunflower sea star. Redundancy at the top of the trophic food web adds resilience and stability to an ecosystem—think wolves and bears in Yellowstone. When sea otters were extirpated in the early 1800s along the Pacific coast, sunflower sea stars not only preyed on sea urchins but, like otters, elicited a behavioral response from urchins to stay hidden. Kelp forests thus maintained viability, but without the presence of otters, they lost resilience. Sea otters have returned to some areas of Alaska and British Columbia, enabling us to study the relationship between these two mesopredators of the kelp forest. Researchers have found that otters and sunflower sea stars go after different urchin. Otters are picky about what they will eat, only choosing urchins full of gonads that offer the caloric benefits they need—that is, the big and juicy ones. Sunflower sea

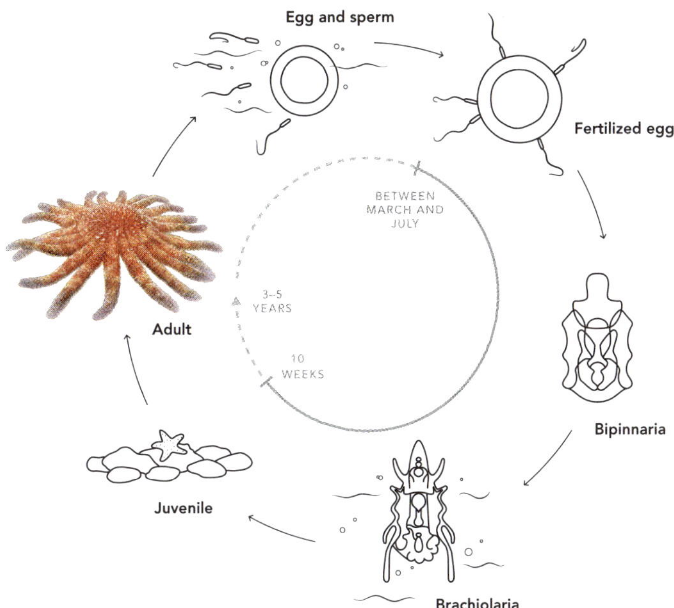

Sunflower sea star life cycle: Sunflower sea stars are broadcast spawners. Each sea star ejects millions of eggs or billions of sperm simultaneously into the water column, often congregating in groups to increase the probability of fertilization. As with urchins, it is impossible to tell a male *Pycnopodia* from a female. The chance fertilization of eggs by sperm results in planktonic larvae that mutate several times over 3–4 months, possibly floating on the currents a great distance from their parent. After ten weeks, the larva settles and grows into a tiny sea star with five legs, and eventually into a 24" wide, 20-armed, colorful, soft-tissued hunter.

stars, on the other hand, will forage for the smaller urchins or the zombie urchins that otters leave behind. Their presence is enough to scare urchins back into their hiding places and allow young bull kelp to grow unscathed. Both otters and sea stars turn out to have an enormous impact on the kelp forest; having two predators at work creates a greater buffer against urchin barrens.

The importance of redundancy was driven home in 2014 when Sea Star Wasting Disease wiped out almost every sea star along the West Coast of North America. While intertidal sea stars recovered quickly, deepwater Pycnopodia are taking their time.

It is estimated that 5.75 billion organisms melted away, leaving them effectively extinct in many locations south of British Columbia, in particular along the California and Oregon coasts.

With kelp forest resiliency already compromised by the extirpation of the sea otter, many bull kelp beds along the coasts of California and Oregon were left with no urchin predator at all. Freed from the threat of the hunter sea star, urchin populations came out of the cracks and started eating down the kelp. Their numbers ballooned. Rising ocean temperatures were also adding stress to bull kelp resilience, and the regime shift from kelp forest to urchin barren was a quick switch in many places. The sunflower sea star's crucial role in maintaining kelp forest health was appreciated only when the sea stars were gone.

Reintroducing sunflower sea stars into the wild seemed wildly farfetched when it was first suggested as part of the developing strategy sessions around bull kelp restoration in the early 2020s. Academic and aquarium marine labs couldn't imitate the sunflower sea star's complex life cycle; the various months-long stages of the young sea star as pelagic larvae seemed impossible to support in the laboratory. But that has changed. The urgency triggered by the bull kelp forest declines and subsequent urchin barrens, especially in California and Oregon, has inspired a raft of researchers to develop the techniques to spawn and raise *Pycnopodia* in tanks on land. Various aquariums, collaborating with experts from a consortium of zoos, marine labs, and academic researchers, have had success, and are now showcasing tank-raised sunflower sea stars in their exhibits. Scientists are analyzing the genetics of these lab-raised stars and noting their feeding habits. Tiny juvenile *Pycno* eat tiny juvenile sea urchins—lots of them!—illustrating an added superpower in controlling urchin densities. Researchers hope that, someday, these young sunflower sea stars will be released into the wild to resume their role as top predators of urchins.

RED ABALONE

Haliotis rufescens

Seven species of abalone live along the Pacific coast, all within the genus *Haliotis*. The red abalone, *Haliotis refuscens*, is the largest and also has the broadest range. The six smaller species—white, black, green, pink, pinto, and flat—are found in more limited sections of the Pacific coast. All six of these are protected in California waters due to overfishing. The pinto abalone fishery has been closed in Washington state, British Columbia, and Alaska since the 1990s.

Range: Baja California to southern Oregon

One of the most alluring residents of the bull kelp forests of the Northern California coast is the red abalone. It holds a special place in the hearts, minds, and mouths of Californians, both native and nonnative. Abalone are naturally abundant around Monterey, California; the word *aulon* traces back to the Indigenous Rumsen people of that area, which then got converted to *abulon* by the Spanish. In Latin, the word for the genus, *Haliotis*, means "sea ear." There are many species of abalone in the world's oceans, all within this same genus.

Abalone is a mollusk, or marine snail, that looks unlike most snails we know. Their spiraled shell is a flattened cap of calcium carbonate that covers the fleshy foot. This foot emerges from the domed shell with a flouncy edge and waving tentacles, and attaches strongly to rocky underwater surfaces. It is a powerful foot force, holding firmly to its substrate. The abalone shell has a row of respiratory holes that run along one edge of the shell, and when found whole or in parts along a rocky shore, the glittering iridescence of the shell's inside, the nacre, makes it impossible to overlook. When building their shells, abalone shift from one polymorph of calcium carbonate to the other, making calcite for the outer shell and then switching abruptly to depositing aragonite for the colorful interior. Light bounces between the layers of aragonite, creating the shimmering iridescence that has been coveted by humans through deep time, a testament to our abiding predilection for beauty. The exterior of the red abalone is brick red, the shells tough and durable, prized for toolmaking; their interior is prized as decoration for masks, sculptures, jewelry, and regalia by California and Pacific Northwest Indian tribes up and down the coast. The earliest fur traders discovered that the easily found abalone shells in Monterey were the best items to use for trade with tribes farther north when bartering for sea otter pelts.

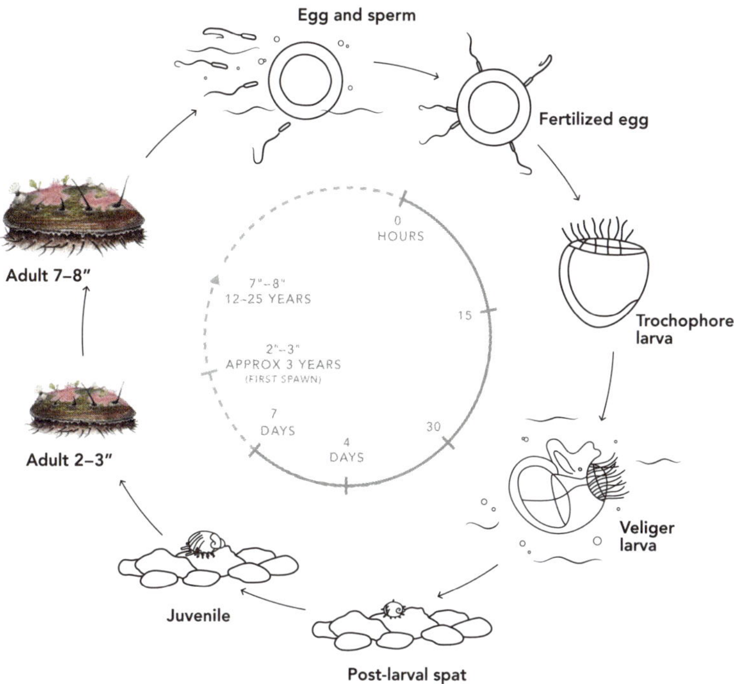

Red abalone life cycle: Male and female abalone broadcast their egg and sperm into the surrounding ocean, where chance fertilization results in planktonic larvae that eventually settle and grow into long-lived abalone.

Abalone, like urchins, are detritivores, living in deep cracks in the rocky ocean terrain or on the sides of boulders, waiting for drift algae or kelp to come their way. In a healthy kelp forest, drift algae are abundant and abalone populations thrive; they are part of the dynamic diversity of organisms there, from fish, marine mammals, sponges, and tunicates to other algae and invertebrates. In fact, an abalone's domed shell provides additional real estate possibilities for a variety of algae and are often seen sporting both encrusting and articulated pink coralline algae, even baby bull kelp. If sea otters are part of that healthy kelp forest—they are voracious predators of urchin *and* abalone—red abalone will stay hunkered in cracks out of reach of the sea otter's grasping paws. If kelp is in short supply, abalone will go looking for it. One of the saddest sights is a photograph of a hungry abalone climbing up the denuded stalk of urchin-ravaged understory kelp.

Red abalone is delicious to eat and was the target of recreational abalone diving on California's north coast for generations, where nostalgia for diving into the kelp forest to pry abalone off the rocks persists. This fishery was closed in 2018 due to the loss of the bull kelp forests that these large herbivores depend on.

ROCKFISH

Sebastes spp.

There are seventy species of *Sebastes*, or rockfish, along the Pacific coast, and more than a hundred species worldwide. They are considered a groundfish and are also known as sea perch, rock cod, or sea bass.

Range: Baja California to Alaska, depending on species

Rockfish is a generic term for a large group of fleshy fish that are delicious to eat—a familiar restaurant menu item. The common names of the species tend to identify them by color; there are yellowtail and yelloweye rockfish; blue, black, and copper rockfish; tiger, canary, red stripe, and silverback rockfish. There are also widow, china, and quillback rockfish. Though their colors vary, they all have a wide, torpedo-shaped body with spiny dorsal fins; large eyes; and an upturned jaw encased in fleshy, downturned lips. To a one, rockfish look grumpy.

Unlike other fish that have distinct territories, many rockfish species live together. In the archipelago of tiny islands of Barkley Sound on the west coast of Vancouver Island, twelve different rockfish species have been identified as residing in that small area. Twenty-eight species are associated with the Salish Sea.

While some rockfish species live in the deeper ocean, rockfish are associated with the kelp forest. This is their home. Their numbers are a sign of the health or decline of their preferred habitat: the nearshore waters in *Nereocystis* and *Macrocystis* territory. Some rockfish spend their entire lives at the same site, making a home among the rocks, seamount cliff faces, and canyons at the ocean bottom. They prefer to stay put, and the sealed air bladder they use to regulate buoyancy is designed for only a gradual trip to the surface. Within a healthy giant kelp or bull kelp forest—with the associated understory kelps of *Pterygophora* and others—phytoplankton, larvae, juvenile fish, smaller schooling fish, and small crustaceans abound. With the added diversity of prey, rockfish tend to eat at a higher trophic level, feeding on small fish, crustaceans, and larvae, who in turn prosper on the algal detritus and nutrients associated with the kelp.

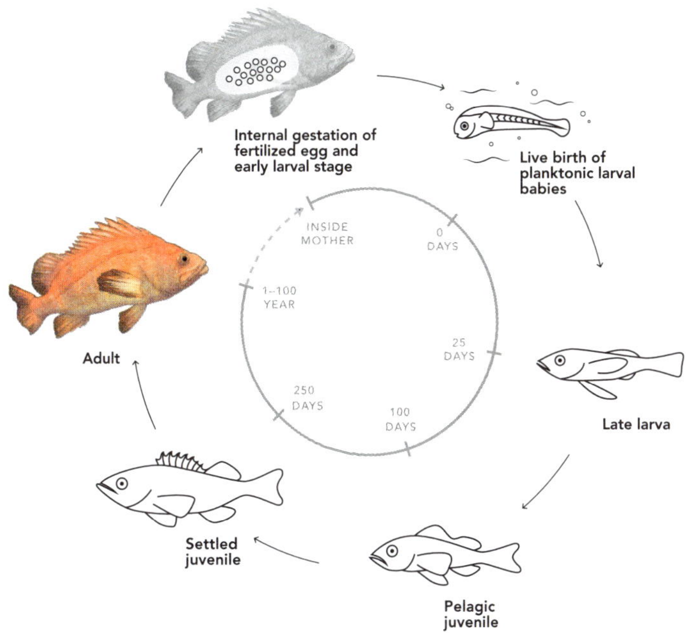

Rockfish life cycle: Rockfish are slow-growing, and some don't start reproducing until 20–25 years old. As they age, the females produce more and more young that are also more robust; they gestate internally for ten months and then birth thousands of late-stage larval rockfish. These larvae are initially pelagic, but then develop into juvenile rockfish in the kelp forests, and then slowly into adults.

Rockfish can be very old. Depending on their species, they can live to be anywhere from 11 to 200 years old. Yes, 200 years old! As a long-lived species, they take many years to reach maturity, not producing young until they are 20–25 years old. The life cycle of the various species varies, but we can look at yelloweye rockfish as an example. These long-lived mothers can carry hundreds of thousands of fertilized eggs, nurturing them internally, and then give birth to live larval young, which float as plankton in the open ocean. A large, 7.5-pound much older rockfish mama can produce 1.7 million young, more than ten times the birth rate of a smaller, younger rockfish. It is essential

to rockfish population health that some of the females reach this mature, bonanza reproductive age.

Humans like to eat rockfish; they are generally mouthwatering and nutritious, but they are particularly susceptible to overfishing. Their site fidelity means that a certain place can be easily fished out. For us to manage fisheries, we need to count fish, but they are hard to count. How can we really know how many rockfish there are in a particular location? Their numbers are positively associated with kelp forests, so as kelp forests decline, so do rockfish populations. Marine Protected Areas, fisheries management plans and regulations, and Rockfish Conservation Areas are important tools for helping rockfish populations maintain resilience under pressure from fishing.

GIANT KELP

Macrocystis pyrifera

Giant kelp is the only species in its genus, but it is a cousin to bull kelp in the Laminaria family. Kelps are a subset of the brown seaweeds that also include rockweeds. All brown seaweeds and kelp have a brown accessory pigment that helps collect light to fuel photosynthesis. The brown pigment mixes with the green chlorophyll to make the giant kelp's golden and sometimes olive color.

Range: Northern Hemisphere—Baja California to Kodiak, Alaska; Southern Hemisphere—Peru south around the southern tip of South America into Argentina

A graphic array of bladders branch continuously off the main stipes of *Macrocystis*, and each bladder sports a luxuriously corrugated blade with delicate serrations along its edge. The pattern of bladders and blades tapers at the tip into a single blade called the apical scimitar. This kelp is the main act of the Monterey Bay Aquarium's central exhibit for good reason. Its pattern, form, and color come together in an organism our human brain recognizes as spectacularly beautiful.

Giant kelp's exuberant growth and sheer beauty are breathtaking both for divers who witness them within the kelp cathedral below the ocean swells and for beach walkers who encounter the swirls of giant kelp washed up on the shore. Like those of *Nereocystis*, giant kelp's contributions to the health of our oceans cannot be overemphasized. *Macrocystis pyrifera* absorbs CO_2 and creates oxygen, creates habitat, and provides protection against coastal erosion. It can reach lengths of 100 feet or more in a single season. Two secrets of the giant kelp's speedy development—up to two feet a day—are that it can grow from any number of points and that it can transfer energy from the top down. The photosynthesizing blades at the ocean's surface distribute their products down the stipe, enabling new fronds to grow farther down the plant. The natural rhythms of the ocean cause the top blades to slough off within months, and the younger blades then rise to the surface for more direct photosynthetic action. The entire giant kelp is a primary production conveyor belt. These multiple growth points mean that the giant kelp can lose, or have lopped off, its topmost vegetation and regrow with abundance—that is, if the oceans are healthy.

Scientists at the Scripps Institute of Oceanography in La Jolla have been studying the kelp forests just off the coast from their laboratories for many decades, while the bull kelp farther

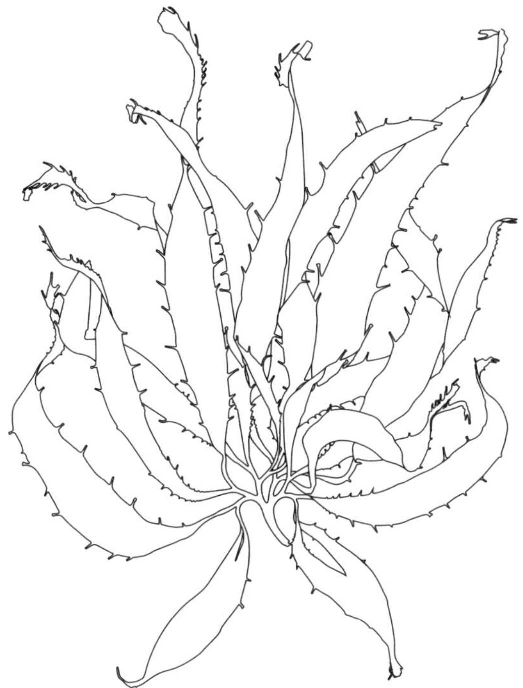

Giant kelp reproductive blades. Specialized sporophyll blades grow at the base of the giant kelp, just above the holdfast, producing the spores that will disperse and germinate into the alternate, microscopic sexual phase of the giant kelp life cycle. Giant kelp is a perennial, however, persisting from year to year, whereas bull kelp is an annual.

north has received less attention, but the destiny of these two fast-growing, massive, and massively important kelps is an intertwined one. Around Big Sur and the Monterey Peninsula, *Macrocystis* and *Nereocystis* do a partnered dance, their ranges intermingling. Generally, bull kelp dominates the kelp forests north of the Golden Gate in California through Alaska and the Aleutians, and giant kelp dominates the warmer waters south of the Golden Gate all the way to Mexico, growing in historically massive kelp beds off of La Jolla and San Pedro near San Diego and Long Beach, respectively. Generally, it is thought that *Macrocystis* likes warmer, calmer waters, inhabiting bays and protected areas of the coast, while bull kelp likes the more dynamic waters and will be found out in the wave-activated subtidal zone, in the rough waters of the cold North Pacific. But these generalities are often turned on their heads: Bull kelp is found in many inland waterways, and giant kelp is regularly found in areas thought to be exclusive bull kelp territory. In Kodiak, Alaska, for example, giant kelp is emerging in hitherto bull kelp–specific regions. Urchin barrens impact both *Nereocystis* and *Macrocystis* beds, but giant kelp restoration work in Southern California, led by the Bay Foundation, is years ahead of bull kelp restoration work farther north.

Competition between bull kelp and giant kelp in a given spot can often come down to their different life cycles. *Macrocystis pyrifera* is a perennial kelp, growing anew each spring from a holdfast that grows and persists through the winter. This is different from its annual cousin, bull kelp, which must grow anew, holdfast and all, each season, and whose tiny young can be shaded out by a perennial giant kelp above it. The giant kelp's holdfast is a magnificent thing, with reproductive sporophyll blades growing out of it at the base of the kelp rather than along the blades above. As it ages, the newer, colorful fingers of the holdfast—they can be red, pink, even bluish—grow over the inner core, making a massive, cone-shaped edifice that becomes an underwater cathedral to hundreds of species that seek refuge in its catacombs. These complex structures are the base of a vertical wonderland of an ecosystem. Understory kelp such as *Pterygophora* grow between the giant kelp, creating an underwater forest of algae that oxygenate the surrounding waters—they are photosynthesizing machines that suck in CO_2 to generate biomass and pump out oxygen at a rate unknown in our terrestrial world. Fish hide among the giant kelp's fronds. Abalone and snails feed on kelp detritus that sloughs off, as do crabs and countless amphipods, isopods, and small crustaceans. Hydroids and bryozoan colonies take up residence on the giant kelp's broad blades and stem, and fish nibble at these delicacies. Urchins are the most voracious kelp herbivore, able to attack the forest in hordes, but also often just one of the thousands of organisms that thrive in the complex system that is the amber forest.

One of the most prized settlers on giant kelp blades is herring roe. Each spring the Haida and other First Nations peoples of British Columbia wait with anticipation for the great schools of herring to lay their roe on the broad blades of the *Macrocystis*. These adorned kelp blades called *kaaw* (pronounced "gow") are valued not only for eating fresh as a delicacy but for drying and trading.

WOODY KELP

Pterygophora californica

Pterygophora californica is known as woody kelp, or stipitate kelp, because its stipe is tough and displays growth rings like a tree when bisected. *Pterygophora* is one of many seaweeds and kelp that make up the understory below the bull kelp canopy. *Egregia menziesii* (feather boa kelp) and *Desmarestia herbacea* (acid kelp) are two other important components of the kelp forest understory.

Range: Baja California to Cook Inlet, Alaska

Pterygophora californica is one of many important understory kelp. Because these robust and bushy seaweeds don't reach the surface or show up on aerial surveys, they are often left out of the kelp forest story, yet they are essential both as habitat and as primary production (bottom of the food chain), much like the brushy chaparral in an oak forest.

Pterygophora has dark-brown blades that are thick and leathery. A perennial, its primary blade grows straight up, and a series of secondary sporophyll blades grow out horizontally from its stipe, like wings. *Pterygophora* has a very stiff stipe up to a meter tall—the stiffest among kelps—and is often referred to as stalked kelp. This stipe grows as a cylinder from the holdfast but flattens up top where the main blade emerges and the wings jut out. When cut in cross-section, the stipe reveals growth rings. In ideal conditions, this kelp can easily live for 15–20 years, and some are 25 years old, an anomaly in the algal realm; it is the oldest kelp in the underwater woods, truly old-growth forest. It tends to grow in two zones, either the deeper kelp forest zone, or closer to shore in groves just beyond the low-tide mark.

Hordes of urchins have swept through many of the bull kelp forests of the California and Oregon coasts since the 2014–2016 ocean heat wave and sea star die-off, devouring all the kelp they can find. On first pass, the urchins eat the bull kelp along with the *Pterygophora* blades, leaving the unbending *Pterygophora* stipes denuded, standing like upright walking sticks. Then, on a second pass, the starving urchins will graze the woody stalks down to nothing. While the bull kelp forest can be picked up on annual surveys and its decline makes the news, the loss of *Pterygophora* is equally alarming.

In the Pacific Northwest and in pockets along the central coast of British Columbia, where the urchin-eating otter have

Pterygophora is a perennial and persists for many years, often in age-group clusters. Sori (spore patches) are produced on the sporophyll blades that jut out horizontally. Baby woody kelp can be found occasionally in the intertidal zone, and like all kelp, they start as a single blade on a single stipe.

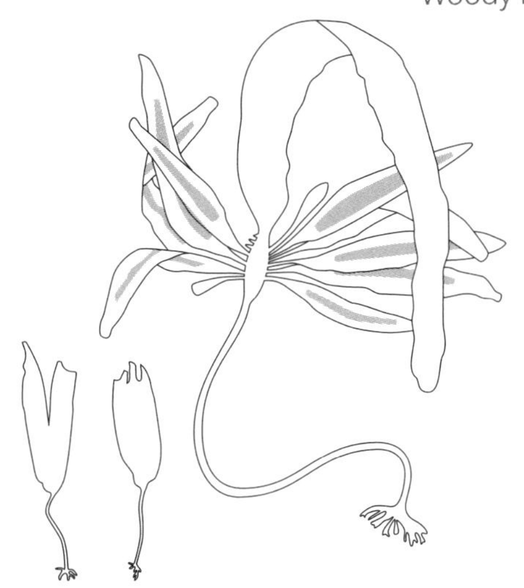

rebounded, *Pterygophora* is abundant. Researchers have studied stands of *Pterygophora* over the long term with dated tags, and they've noticed that whole groves will senesce, or age out, at the same time, and wash ashore in great piles of stipes, devoid of blades, along the wrack line. They turn a deep orange and then dry to black. Easily mistaken for sticks or pieces of driftwood, the stipes have flattened ends and corrugated, regularly curved edges—telltale signs that they are algal in origin. Eagles and osprey favor these hardened stipes for their nests.

Pterygophora has a few common names, winged or woody kelp among them, but the most poetic name, used by Indigenous people of the Pacific Northwest, is "walking kelp." Its holdfast can attach to a cobble or a stone heavy enough to keep the organism rooted to the ocean bottom, but as the kelp grows, vulnerable to wave action and surf, the cobbles, with their attached kelp, bounce in the surf, creating the effect of "walking" around the shallower waters. First Nations peoples would use this mobility to their advantage by arranging the *Pterygophora*—with their weighted cobbles—into rows, creating a funnel to deflect salmon into fishing traps. Ingenious!

SNOWY PLOVER

Anarhynchus nivosus

Snowy plovers are one of twenty-four species of plovers, including the piping plover and the semipalmated plover, but the snowy plovers on the West Coast are actually members of the subspecies *Anarhynchus nivosus nivosus*, or western snowy plover. Snowy plovers along the Pacific coast have been listed as threatened under the Endangered Species Act since 1993.

Range: Coastal beaches from Baja California to northern Oregon

Snowy plovers are tiny—an ephemeral puff of a shorebird. They are no bigger than 6 or 7 inches long and weigh only 1–2 ounces. Despite their distinctive dark markings at the eye and throat, their white underbelly and sandy-brown top plumage make them almost invisible on the open beach where they live. Their short, spindly legs motor them across the sand in a constant stop-and-go foraging quest for bugs. Their large eyes indicate their foraging behavior as visual predators, snatching at bugs on the sand with their strong beak.

Snowy plovers, unlike other shorebirds, do not migrate up to the Arctic in spring. They stick around the coastal temperate zone of the Pacific coast, making their home in the upper dunes of sandy beaches from Baja through the great sandy stretches of Oregon. This vulnerable yet resilient powerhouse of a shorebird

makes its home in spaces we humans covet, and its biogeography maps directly onto heavily anthropocentric beach zones. They prefer to breed in undisturbed dunes with native diversity of flora and plenty of kelp wrack, not only for food provisions but also for creating the smallest of bulwarks on the sand to lay some eggs. An elegant curve or crook in a piece of bull kelp, with sand blown up against it, makes the perfect rampart for the snowy plover's nest.

The female lays two to six eggs, expending enormous energy to lay and brood the eggs. The eggs hatch simultaneously after twenty-six days in what can seem like barely a divot in the sand. The tiny fuzzball hatchlings immediately start hunting and eating insects themselves, returning to the father at the nest, who takes over brooding the young for about four weeks. And so the snowy plover's world is intimately tied to the nuances of topography of the upper beach, where the beneficence of the fresh wrack gives way to the structural benefits of dried bull kelp stipes, which hide barely discernible nests and plover babies. In the four weeks before hatchlings fledge, they grow eightfold in biomass, all due to the constant supply of insect protein, the trophic subsidy brought ashore by the bull kelp and other algae from the nutrient-dense waters of the Pacific Ocean. Nest predation by ravens and crows, who have adapted so effectively to human environs and become the most successful egg thieves of all, is one of the greatest threats to plover populations.

PELAGIC CORMORANT

Urile pelagicus

The pelagic cormorant is one of the three cormorant species of the Pacific coast, alongside the Brandt's cormorant and the double-breasted cormorant.

Range: Baja Pacific Islands in Mexico up the coast to the Aleutian Islands

The pelagic cormorant is a good example of a fish-eating seabird that benefits directly from the kelp forest as habitat engineer. The bull kelp forest is home to anchovies, herring, and up to thirty different kinds of young and forage fish (also known as prey fish or bait fish) that cormorants eat. Cormorants are spectacular divers, becoming black torpedoes underwater, and can hold their breath for up to two minutes, preying on fish within the bull kelp forest.

Everything about a cormorant's stature and look is in the service of this diving enterprise. Pelagic cormorants are slender birds, the smallest of the three Pacific cormorants, standing 20–30 inches tall. Their adult plumage is very dark, but can shine iridescent green in sunlight. The only spot of color appears in breeding adults, as a patch of white behind the wings and a spot of red at the edge of the mouth. A cormorant's bones are solid and their feathers become waterlogged; both of these characteristics, along with the bird's webbed feet, are adaptations for efficient diving. You can often see cormorants perched on a rock, extending their soggy wings outward to dry.

Despite its name referencing the open ocean, the pelagic cormorant hugs the shores of the North Pacific, rarely flying farther from the coast than the strip of ocean associated with the rocky kelp forest. The three species of cormorant found along the Pacific coast—pelagic, Brandts, and double-crested—take care not to compete with one another, foraging in slightly different zones. They also nest in different ways. Pelagic cormorants nest on small slivers of shelf on cliffs facing the ocean, laying eggs and growing young in what seem like impossible circumstances. A family of five or six will be crammed onto a tiny ledge. These cormorants use kelp wrack to build their nests along coastal cliffs, using guano to stick the kelp and seaweed fronds together and

glue them to the steep rocks, adding stability to the precarious positioning. It's not uncommon to see cormorants stealing seaweed from one another.

By late July or early August, each nest might have two or three young cormorants, as big as their parents but identified by their lackluster feathers that are not yet shiny. Their speedy growth, from fuzzy chick to full-sized cormorant in two months, is a testament to the nutrient riches offshore brought back to the nest by the parent cormorants and dispersed to hungry young on the cliffside. The diversity of fish that live and grow in the bull kelp forests is a key factor in these chicks' rapid development.

The year-round range of the pelagic cormorant closely mimics the range of bull kelp itself; the success of its breeding populations, countable by shore-based observers with good binoculars, is a clear indication of the health of the kelp forests we cannot see.

MARBLED MURRELET

Brachyramphus marmoratus

Marbled murrelets are in the Alcidae family of birds, related to auks, murres, and puffins.

Range: Central California north through the Pacific Northwest to Alaska and the Aleutians in those places where there is old-growth forest in proximity to the coast

The marbled murrelet is a stocky seabird that is, perhaps, the most literal connector between the forests below and the forests above. It is mottled brown with a short, thin bill that points upward when the bird is swimming on the ocean surface. It is an acrobatic diver, catching sardines, herring, squid, and zooplankton in the bull kelp forests of the North Pacific, but nesting and raising its young in old-growth trees of the adjacent coastal forests. In today's world, with limited old-growth forests left in the lower forty-eight states of the US, the marbled murrelet is most commonly found along the foggy coastlines and rocky waters of Alaska and northern British Columbia. Unlike cormorants, it uses its wings underwater, diving quickly to catch small fish relatively close to shore. It brings larger fish—anchovies and herring—to its nesting young, flying at great speeds and often long distances, to transfer these nutrients from the ocean environment to the forests. It builds nests on the long horizontal branches covered in moss and lichen of old Sitka spruce, Douglas-fir, Alaska yellow cedar, western red cedar, western hemlock, mountain hemlock, and coast redwood, most often choosing the tallest tree in the area. But "building" is perhaps too suggestive, as there is no nest built: The marbled murrelet lays a single egg in a depression in the moss on a large branch in the forest canopy, and a single chick is hatched and fledged.

Marbled murrelets are elusive birds, making their connection between ocean and forest even more poetic. Their nesting habits were a mystery for years because their old-growth habitat—such as the redwood forests of Northern California—had been destroyed in most places where people were trying to find them. Their nests were only discovered by happenstance in 1974. Although their nests and young are raised in the forest canopy, the courtship of two murrelets begins at sea in late winter and continues into the spring. It is an ocean affair. Partners swim side by side, point their bills skyward and then dive and surface simultaneously. Pairs chase each other in flight, then suddenly continue the pursuit underwater. It is an aquatic love story, the pair monogamous throughout the breeding season.

PINK SALMON

Oncorhynchus gorbuscha

Pink salmon are one of five species of Pacific salmon: The largest is Chinook, or king salmon (*Oncorhynchus tshawytscha*); the smallest is pink salmon, or humpies (*Oncorhynchus gorbuscha*). The others are sockeye, coho, and chum salmon. Steelhead and cutthroat trout are also in the *Oncorhynchus* genus, so can also be considered salmon.

Range: Straits of Juan de Fuca north through British Columbia and the waters of Alaska

Pink salmon are not pink. They are silvery gray, slightly rounded, and the smallest of Pacific salmon, so they travel in schools for safety in numbers. Fishers are respectful of the small but mighty pink salmon; they do not succumb to the lure easily. It is their flesh that is pink—not orange—giving them their name, and it is less oily and milder in flavor than that of other salmon.

Pinks have the shortest life cycle of the salmon, living only two years, and they quickly leave their birth rivers through estuaries and swim out into the ocean. As schools of young pink salmon emerge into the ocean, they are vulnerable to predators of all sizes, including other salmon, so the protection of the nearby kelp forest is essential. When pinks return to their home rivers to breed, they develop spots on their backs and tails, and on the last leg of their journey just before they spawn, males develop a large hump behind their head and the tell-tale hooked jaw.

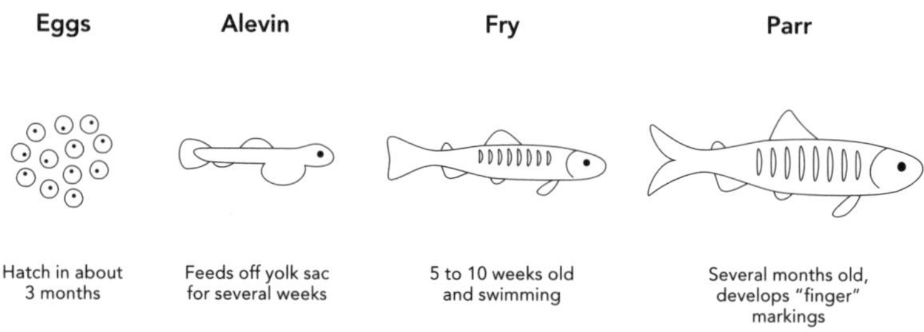

Pink salmon, like all salmon, are anadromous—they transition from fresh water to salt water—and their transition into the marine environment is where the kelp forest and salmon stories intertwine. Bull kelp find refuge in the ocean waters adjacent to river openings and tend to have healthy beds in those areas. This gives young salmon emerging from the rivers a chance to hang out in the three-dimensional habitat of the kelp forest and adapt to the ocean. The kelp forest is a space of biodiversity, so whether the young salmon's food source is krill (euphausiids), shrimp, or smaller fish, the kelp forest is the place to find it. This kelp–salmon story is being written by scientists right now. Although there are many visual accounts, there is very little data to quantify this relationship. Those who study and dive in the bull kelp forest; who see a seal break the surface from the kelp beds shaking a pink or orange blur, a salmon, in its jaw; who count salmon prey and gauge fisheries' health, know that the bull kelp forests provide habitat and migratory protection for salmon. But recording this kelp–salmon connection is another challenge. Divers and snorkelers on the scene are sure to skew the number of fish in the vicinity.

Alaska is ground zero for this new science: Salmon/kelp surveys are recording salmonids of all kinds and ages in the wild bull kelp beds along the southern coast, but also in the kelp

The salmon life cycle has seven named stages. Each species spends a different amount of time in each stage, but every salmon adapts from fresh water to salt water in the ocean; it does the reverse when it returns to its native river or stream to spawn, and then dies.

farms now common around Kodiak, Juneau, and Prince William Sound. While the methodologies are still being developed, anecdotes abound: Fish and kelp go together; the data is coming. First Nations peoples know, researchers know, and fisheries managers know that healthy bull kelp and healthy salmon are vital parts of the symphony that is the healthy kelp forest.

Pink salmon are the most abundant of the wild Pacific salmon, and if fished and processed properly for consumption, they can be an important contributor to the long-term health of salmon populations, taking pressure off the more iconic, larger, but more endangered Chinook and sockeye salmons. Around Lummi Island, in the Salish Sea, traditional reef net fishing—where a watcher standing high up on a platform eyes the groups of silvery salmon as they pass over a net—continues, catching pink salmon schooling with the tidal currents in the narrow passageways around the island. The Lummi tribe and other fishers of the area are dedicated to keeping traditions alive and promoting the status of pink salmon as desirable and delectable for market. This species is the future of salmon fisheries.

HUMANS

Homo sapiens

Humans, or *homo sapiens*, are us—we. We are the only species in the genus *Homo*.

Range: Humans inhabit all the continents of Earth except Antarctica. Many human groups have gravitated toward living along coastlines adjacent to the oceans.

Bull kelp evolved well before humans arrived, under pressure from prehistoric grazers as well as in response to the nutrient loading of nearshore waters by the guano of ancient bird colonies. But ever since humans arrived, they have been part of the ecological mix of each coastal region along the eastern edge of the Pacific. In 1975, archaeologists discovered remains of a 12,000-year-old human settlement at Monte Verde in Southern Chile. Among the artifacts preserved in the peat bogs were the clear remains of nine species of marine algae from distant beaches and estuaries. These seaweed relics suggest that the kelp beds along the west coasts of North and South America created a nutrition highway—the kelp highway—along which the first humans traveled either by boat or along the shore, eating the abundant shellfish and seaweed there. Oysters, clams, seaweed, crab, and fish from the shallow ocean estuaries and beaches of the time (many miles out from today's shoreline) provided the fatty acids (important omega-3s), iodine, and other minerals essential for human development. The complex human brain requires a very specific mix of nutrients found only in these resources. Iodine, in particular, gave our evolving brain the boost it needed to create language and fully realized societies.

For millennia before contact, Indigenous and First Nations peoples living in the territories adjacent to bull kelp beds, many of them oceangoing cultures, used bull kelp bladders as vessels; picked kelp to eat; used its long, thin stipe as fishing line; and integrated it into art and toys. For example, Haida and Samish stories acknowledge the beneficence of the bull kelp forest for sustaining salmon and other fish, abalone, and urchin—the ocean bounty their cultures flourished amid. Humans competed with sea otters as top predators of this bounty. Indigenous practices of selectively culling sea otters are now acknowledged and have even been reinstated in Alaska to protect shellfish for human sustenance. The sea otter's luxuriant fur was traditionally used in garments and regalia, and is now being used again by Native Alaskans.

Nonnatives have much to learn from Indigenous peoples' intimate understanding of this landscape, which is reflected in their languages. This is in contrast to the hierarchical, artificially differentiated taxonomic thinking that pervades Western science and sets humans aside from nature. Hawk Rosales, an Indigenous land defender of Ndéh ancestry, writes:

> *Languages foreign to the lands and waters of this hemisphere do not contain the words, ideas, or even the sounds needed for conveying the depth and intricacy of Indigenous understandings of the natural world. Central to the place-based cultural lifeways of Indigenous Peoples are our languages. Our languages enshrine and order the unique attributes of these places and their myriad interactions, and the realities within which they exist—both physical and metaphysical. Like the lands, waters, and skies, Indigenous languages are supremely beautiful and powerful. They derive from and describe origins, behaviors, and relationships among the many beings and elements of specific places. These places include the animal, plant, human, and spirit relations who have dwelt in them since time immemorial.*

As Indigenous ecologists perform more of the naming work in their own languages, science and culture come together to connect communities and broader audiences to the kelp forest in different ways.

Throughout this book, we have mentioned many ways that humans have impacted the kelp forest and environs here on the

West Coast of the United States: the extirpation of the sea otter by nineteenth-century hunters, the resulting shellfish bonanza and markets centered on abalone and red urchin, overfished rockfish populations, and real estate's incursion into shorebird habitats—to name only a few. We have also gestured toward the role of science, how our knowledge of our nearshore ocean community has evolved alongside these market forces and informed ecological interventions. For example, shellfish and fish are directly beneficial to humans as protein-rich food and thus gained commercial value, attention, and study by humans. Bull kelp, by contrast, despite being the foundational habitat, did not have commercial value, so its particular study and management were more obscure. Only when the kelp forests of Northern California, Oregon, and Puget Sound, and then the Salish Sea in British Columbia, began to disappear did humans take notice. A decade ago, humans created a system of Marine Protected Areas (MPAs, where human activity is limited) as one strategy for addressing the loss of biodiversity in the changing oceans. The idea was to create refuges for ocean organisms, free from fishing and other human interference. The first decadal surveys of MPAs suggest a modest positive outcome.

The loss of bull kelp, the prospect of warmer oceans, and the intransigence of sea urchins to die off naturally have prompted a much more interactive human approach. It has become clear that these depleted urchin-barren oceanscapes could become the "new normal" if humans don't get involved. Ecologists are creating kelp restoration techniques in real time as the crisis is under way. It is new and experimental work, requiring whole years to iterate, given bull kelp's annual life cycle. It might entail urchin divers culling urchins from specified kelp recovery sites so that baby kelp can grow without getting eaten; outplanting nursery-grown baby kelp; or seeding a former kelp bed with rocks inoculated with kelp spores. Agricultural and forestry practices on land have been developed over centuries, but there has not been any such development in the cold and inhospitable waters of the North Pacific. The ocean is a hard place for humans to work, and the bull kelp forests hold many secrets still to be understood. Collaboration and cooperation across sectors and regions are key to success. We humans are trying to learn from the supportive web of the kelp forest itself.

Bull Kelp

REGIONAL STORIES

Bull kelp grows in a narrow band of the Pacific Ocean hugging the coast from Central California to the Aleutian Islands, Alaska.

We have chosen eight regions to describe the kelp forest drama that is playing out under the lid of the tide, out of sight by all but accomplished free divers, scuba, and commercial urchin divers. These eight stories illustrate a vast range of outcomes, from healthy, diverse, wild kelp beds to kelp forests transformed into wastelands of spiky urchin barrens. Comparisons of one region to another enable us to understand how such factors as ocean temperature and predator (sea otter and sunflower sea star) presence or absence can affect the foundational kelp forest habitat so profoundly.

We have created a bespoke bathymetric chart—a map showing the contour of the ocean floor—for each region. Bull kelp needs prodigious amounts of daylight and ocean nutrients to grow the 6–10 inches a day that it does in springtime. It inhabits the zone labeled on the charts as the sunlight zone, nearest to the continent's edge. Each region's bathymetry implies a different range of conditions that lead to unique kelp forest ecologies.

outheast Alaska
British Columbia
Puget Sound,
Washington
Washington Coast
Oregon Coast
Northern California
Central California

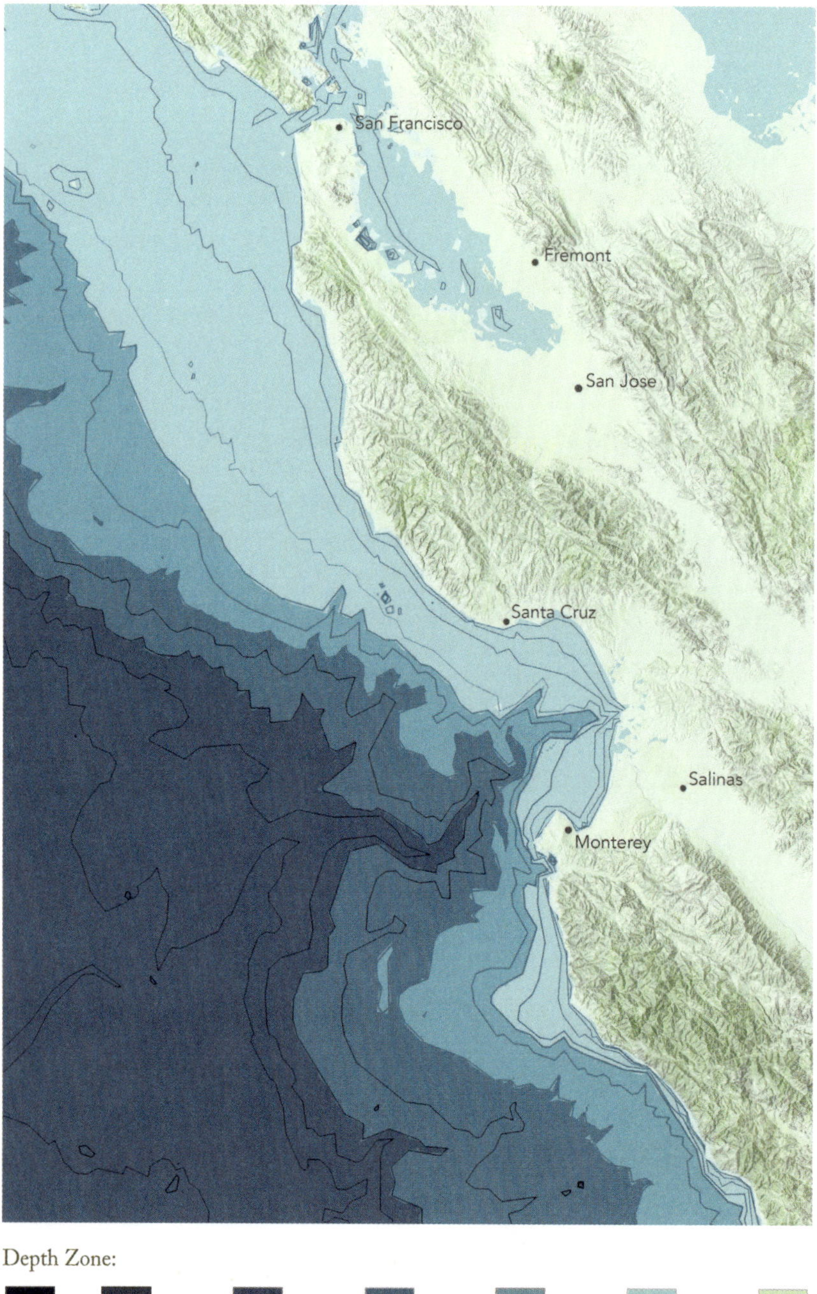

San Francisco
Fremont
San Jose
Santa Cruz
Salinas
Monterey

Depth Zone:

Central California Coast/Monterey

> *As seasons progress in the Monterey Bay kelp forests, residents adjust to the sunny calm of spring or the violent storms of winter. Each year brings different patterns of weather, disease, and competition for food and space in the forest. Sometimes conditions are poor, but from time to time, when conditions are right, it's like an enchanted forest.*
>
> —Judith Connor, retired Monterey Bay Aquarium Research Institute biologist

The great Monterey Canyon stretches from the deep ocean toward the center of Monterey Bay like a dragon, its snout dissolving into the Elkhorn Slough and its leg stretching along the southern side of the Monterey Peninsula, with claws extending into Carmel Bay. The magnificence of this underwater canyon is visible on bathymetric charts, but it is otherwise invisible to us shore huggers. The effects of the constant mixing of deepwater nutrients from the canyon into shallower zones are clear as soon as you venture to the shore and experience the intense richness of the Monterey Peninsula tide pools or get a chance to snorkel, dive, or kayak just offshore. The abundance and variety of species of seaweed are staggering! These are some of the most biodiverse waters on earth.

Are they here?

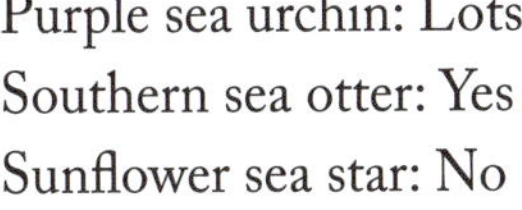

Bull kelp: Patchy, intermixed with giant kelp
Purple sea urchin: Lots
Southern sea otter: Yes
Sunflower sea star: No
June 2025 average sea-surface temperature at Monterey: 13°C / 55.5°F

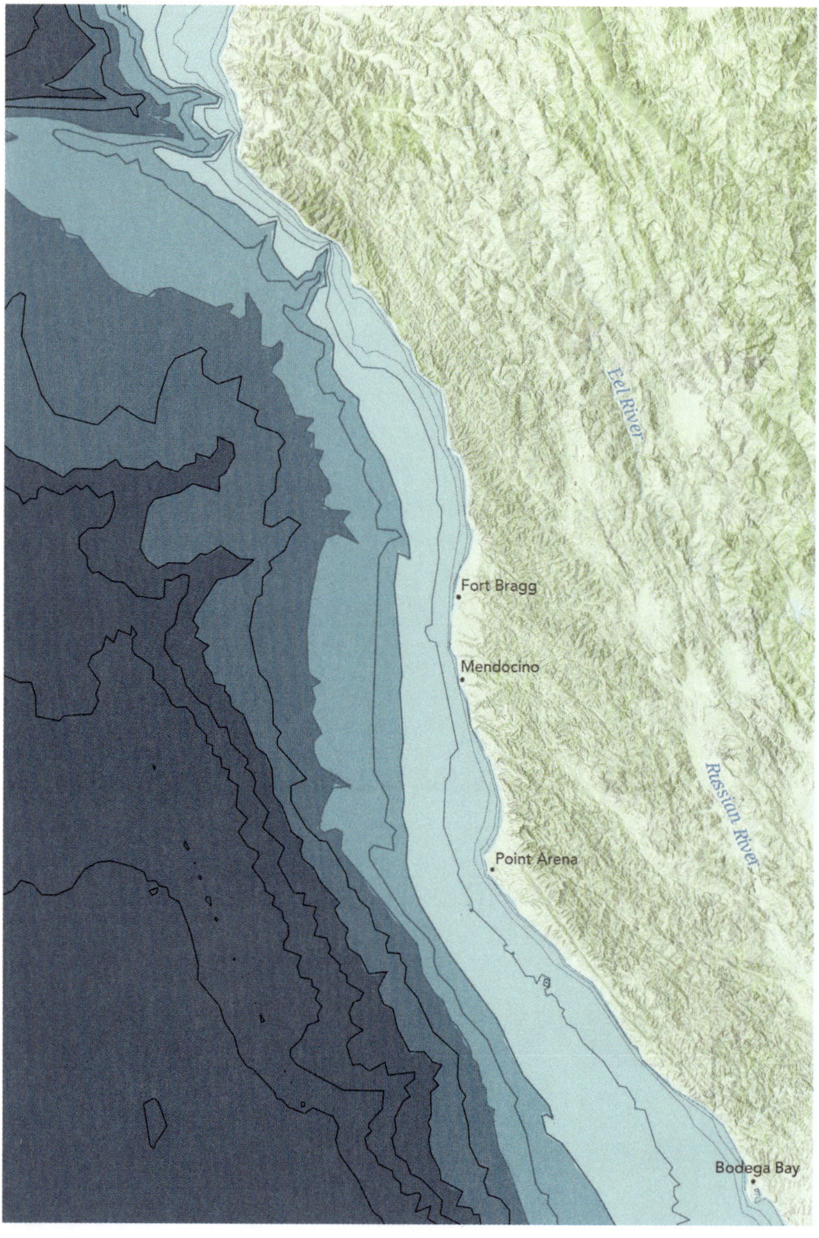

Eel River

Fort Bragg

Mendocino

Russian River

Point Arena

Bodega Bay

Depth Zone:

Northern California/Mendocino County

> *As more time goes by, it's easy to forget the richness and the fullness of a thick and healthy bull kelp forest along the Mendocino coast. It has been almost ten years since the ocean warming events that changed everything about the way the flora and fauna exist in a kelp forest here. The ecosystem has yet to recover on its own.*
>
> —Tristin Anoush McHugh, The Nature Conservancy

There are no sea otters along this rugged coast. Sea otters have not swum regularly in Northern California waters since they were extirpated in the fur trade by the 1890s. Without predation by sea otters, red abalone grew large and abundant, and recreational abalone diving (no scuba, just free diving allowed) was a favorite activity. The abalone diving community produced economic opportunity in former mill towns along the Mendocino coast, and it created a culture of connection to the coast, to family, to place. Until recently, sea otters were reviled by fishers and abalone divers alike as "rats of the sea," competing for abalone and commercially valuable red urchins. Now that the kelp has declined and urchins dominate the ocean bottom, the abalone diving fishery is closed, and sea otters are acknowledged as an important part of a healthy kelp forest system. Unfortunately, they cannot survive in an urchin barren. The urchins are empty; they do not hold the rich uni that sea otters need to fuel their hypercharged metabolism.

Are they here?

Bull kelp: 94% decline from 2013
Purple sea urchin: Lots. Urchin barrens are common.
Southern sea otter: No
Sunflower sea star: No
June 2025 average sea-surface temperature at Fort Bragg: 11°C / 52°F

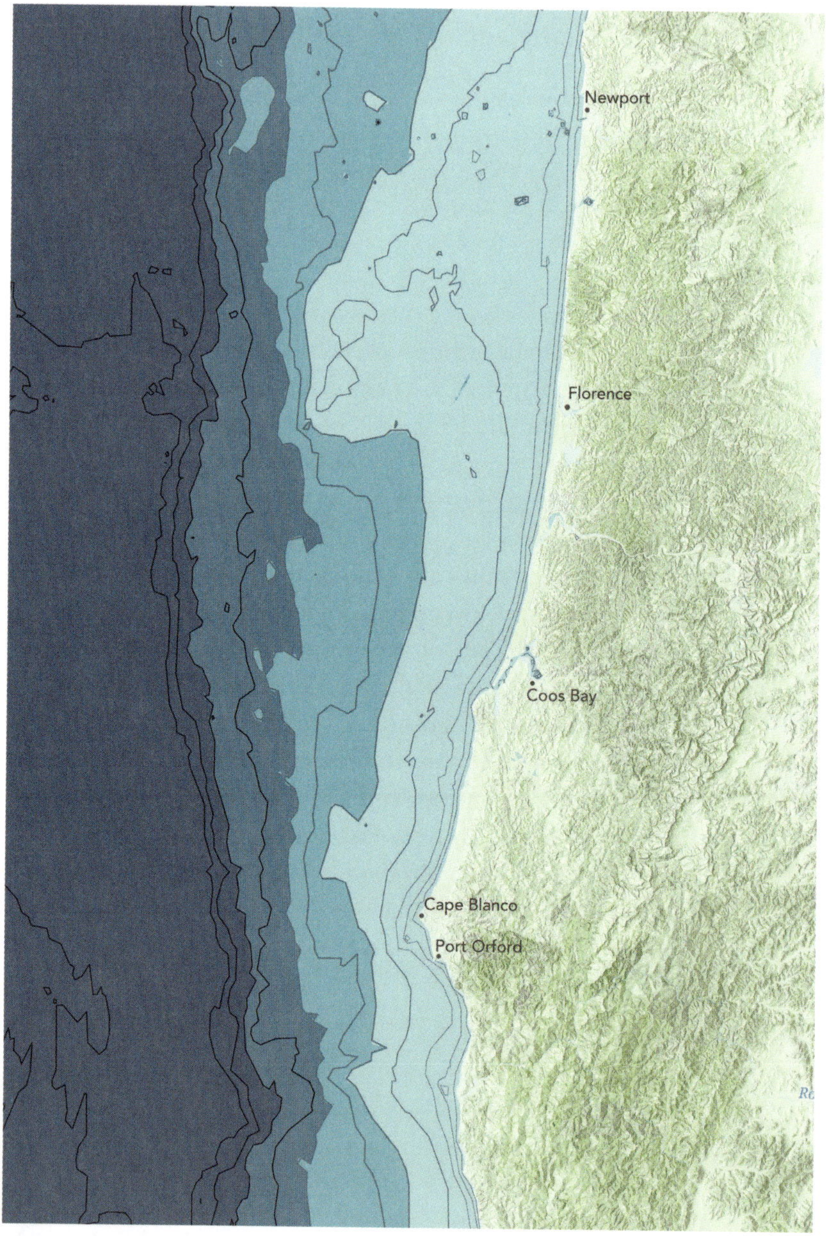

Newport

Florence

Coos Bay

Cape Blanco

Port Orford

Depth Zone:

Oregon Coast

Our coast is a story of contrasts.

— Tom Calvanese, Oregon Kelp Alliance

The Oregon coast is full of contradictions: It has amazing rocky reefs, perfect bull kelp habitat, and long expanses of sandy dunes and beaches, not kelp habitat at all. Its coast is stunning and wild and full of life, yet only a tiny portion of it, less than 2 percent, is protected in five distinct marine reserves surrounded by Marine Protected Areas, positioned almost equidistant from each other along the seaboard. The salty Pacific Ocean is interrupted by the outflow of seven great rivers along this Oregon stretch, bringing enormous quantities of fresh water to mix with the salt. Oregon's fisheries are intense commercial enterprises, pulling crab, rockfish, and urchin out of the coastal waters, and the fishers are often the frontline observers of ocean change. Though there are two coastal landmarks named for sea otters, Otter Rock and Otter Point, there are no sea otters on the Oregon coast. Around Port Orford in 2024, there were sustained kelp beds at Rogue River Reef, but urchin barrens just to the north at Orford Reef.

Are they here?

Bull kelp: Patchy, with areas of much loss
Purple sea urchin: Lots. Urchin barrens are common.
Southern sea otter: No
Sunflower sea star: No
June 2025 average sea-surface temperature at Port Orford: 11°C / 52°F

Tatoosh Island
Neah Bay
Strait of Juan de Fuca
Ozette
La Push
Olympic
Peninsula

Depth Zone:

Trenches
(>6,000 m)
Abyss
(4,000 m–6,000 m)
Midnight Zone
(2,000 m–4,000 m)
Midnight Zone
(1,000 m–2,000 m)
Twilight Zone
(200 m–1,000 m)
Sunlight Zone
(<200 m)
Land

Washington Coast

I want the sea. That is my country.

—Makah Chief Tse-kaw-wootl from Ozette to Washington governor Isaac Stevens, during the negotiations of the 1855 Treaty of Neah Bay

Tatoosh Island is a hiccup off Cape Flattery, the pointy nose of the Olympic Peninsula. Tatoosh is a tiny island with enormous importance to the Makah, the Cape People, as a whaling station and as central to their territory and lifeways along this section of the coast of Washington state—an amalgam of pounding surf; high, spiked sea stacks; and sandy beaches. The land turns a sharp corner here, tracing the edge of the dynamic, open Pacific Ocean into the quieter Strait of Juan de Fuca. The bull kelp beds along this section of coastline, both facing the open Pacific and into the Strait, are historic, robust, and healthy. They offer us humans ample lessons to learn from.

Are they here?

Bull kelp: Healthy
Purple sea urchin: Some
Northern sea otter: Yes, plenty, some living in large rafts
Sunflower sea star: No
June 2025 average sea-surface temperature at Neah Bay: 11.5°C / 53°F

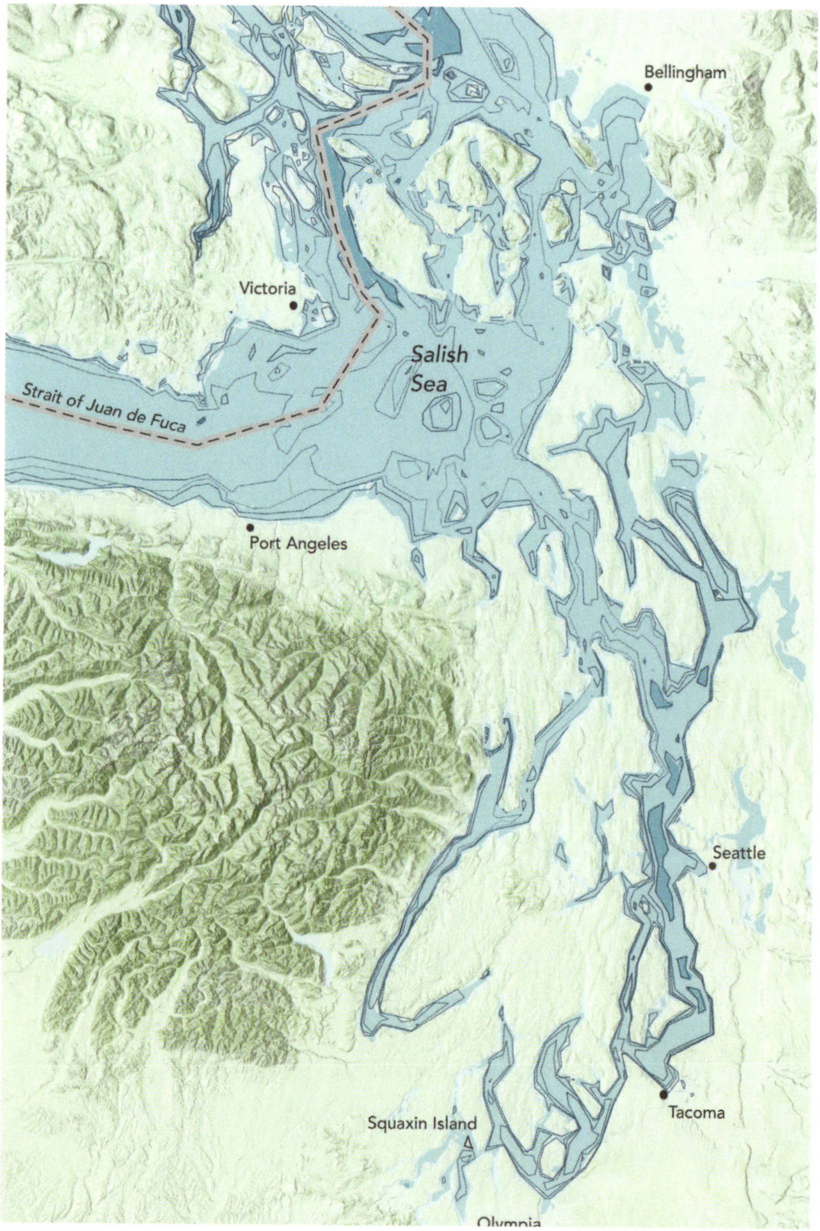
Bellingham
Victoria
Salish Sea
Strait of Juan de Fuca
Port Angeles
Seattle
Tacoma
Squaxin Island
Olympia

Depth Zone:

Trenches (>6,000 m)
Abyss (4,000 m–6,000 m)
Midnight Zone (2,000 m–4,000 m)
Midnight Zone (1,000 m–2,000 m)
Twilight Zone (200 m–1,000 m)
Sunlight Zone (<200 m)
Land

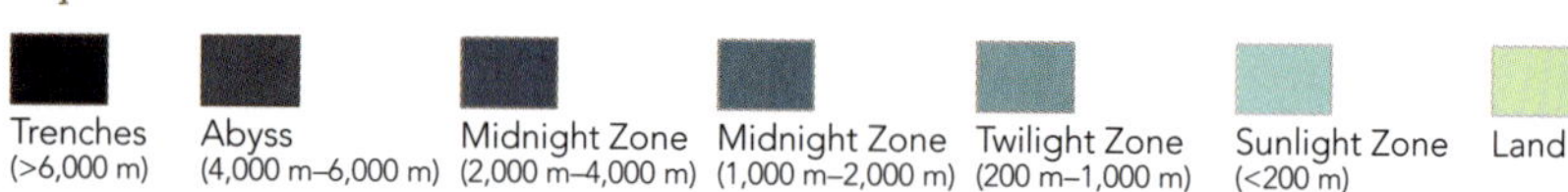

Puget Sound, Washington

> *Ever since humans followed the kelp highway to reach these shores, our lives have been interwoven with kelp. We have navigated by it, found safe harbor in it, eaten it, used it in cultural practices, and harvested the animals great and small that live in it. Now we need to look to it for our future.*
>
> —Betsy Peabody, Puget Sound Restoration Fund

Puget Sound is a long, thin appendix of water fed by the twice-daily tidal flux of the Pacific Ocean coming into it through the Strait of Juan de Fuca, and by the six great rivers and estimated 2,800 streams of its watershed. This great mingling of salt water and fresh water is mediated by two underwater sills, or humps, that create basins within Puget Sound, which act as mixing bowls creating turbulent currents and flows. Despite its distance from the ocean, this flow of current and mixing is what has made Puget Sound prime bull kelp habitat since the Sound was scoured by the last ice age and the glaciers retreated 10,000 years ago. The bull kelp forests of Puget Sound have provided habitat and a detrital food source for the rich fish and marine life for which Puget Sound is famous. The salmon, rockfish, crab, clams, and oysters that depend on the kelp forest have sustained generation after generation of those humans living on the Sound's adjacent shores. But these waters are warming, and bull kelp beds are greatly reduced. And while Puget Sound does not have an urchin issue, it does have a kelp crab issue. Too many of these crabs eat and claw at the bull kelp, further eroding its resilience.

Are they here?

Bull kelp: Historic beds are greatly reduced.
Purple sea urchin: A few
Kelp crabs: Lots
Northern sea otter: None
Sunflower sea star: None until you get north into the Salish Sea where a few have been spotted
June 2025 average sea-surface temperature at Seattle: 13.5°C / 56°F

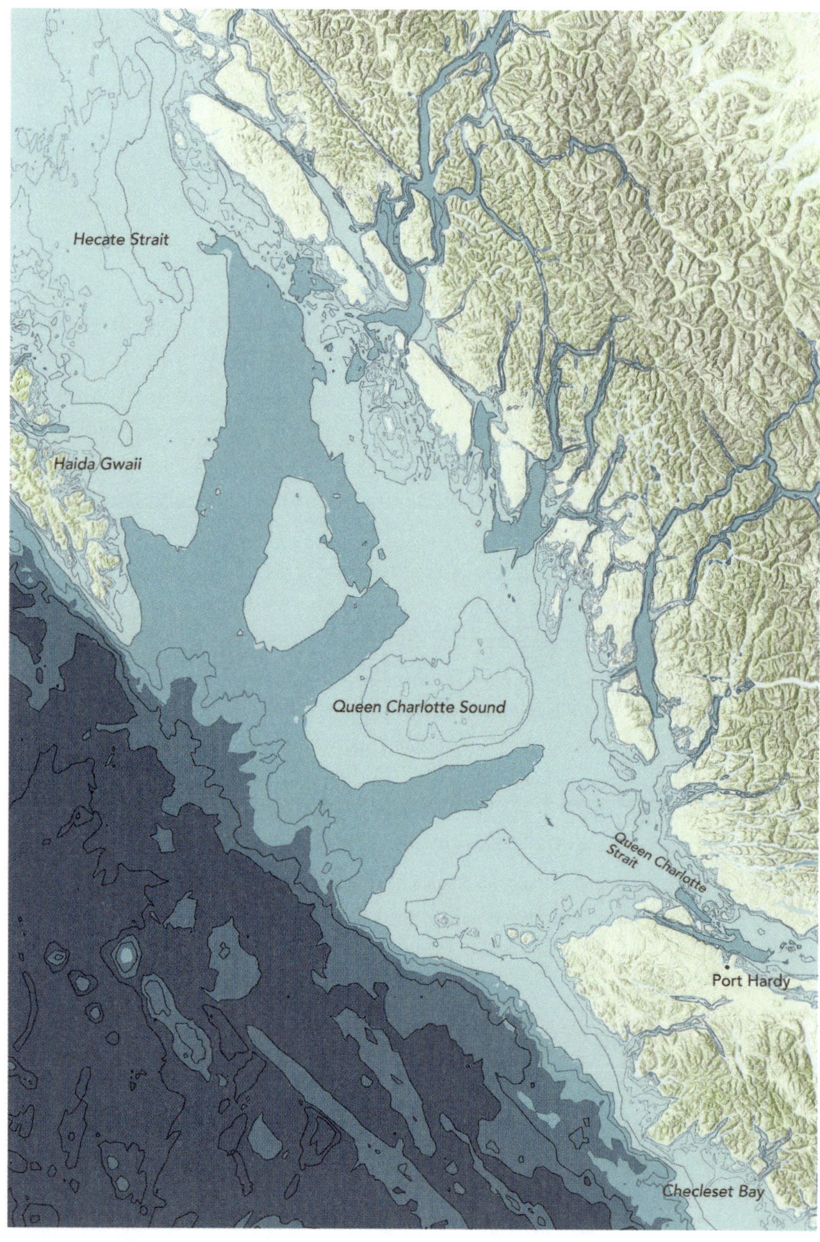

Depth Zone:

Trenches (>6,000 m)	Abyss (4,000 m–6,000 m)	Midnight Zone (2,000 m–4,000 m)	Midnight Zone (1,000 m–2,000 m)	Twilight Zone (200 m–1,000 m)	Sunlight Zone (<200 m)	Land

British Columbia, Canada

> *The diversity of kelp on BC's vast coastline is unmatched and provides unique opportunities for research and restoration.*
>
> —KelpRescue.org

The BC coastline's interaction with the Pacific Ocean is endlessly complex. The jagged outer coast of Vancouver Island directly faces the wide Pacific, while its inner shore flanks a web of narrow waterways and straits running between the countless islands that connect the Salish Sea to the archipelago of Haida Gwaii. The Pacific infiltrates the mainland edge of British Columbia with deep fjords and currents running back and forth around thousands of marine islands. If the BC coastline were stretched out straight, it would extend for 15,985 miles, or 25,725 kilometers. The magnitude of this varied algal habitat is mind-boggling. These are bull kelp waters, with cold ocean churning with waves or strong current. Bull kelp is a flow hound—it loves the small passageways between islands where the Pacific is forced through first one direction and then the other, bringing fresh nutrients with every turning tide. Its exuberant growth quickly uses up the nutrients in surrounding waters, so the influx of new, nutrient-dense ocean water is essential for its success through the season. On the open coast, this churning of nutrients might come from dynamic wave action. In the interior waterways, it comes from currents squeezing through deep channels between land masses with rocky shores.

Are they here?

Bull kelp: Healthy in some places, in decline in others

Purple and green sea urchin: Some. This is the transition zone from purple to green sea urchin.

Northern sea otter: Repopulating certain areas of outer Vancouver Island and the mainland

Sunflower sea star: A few

June 2025 average sea-surface temerature at Port Hardy: 11°C / 52°F

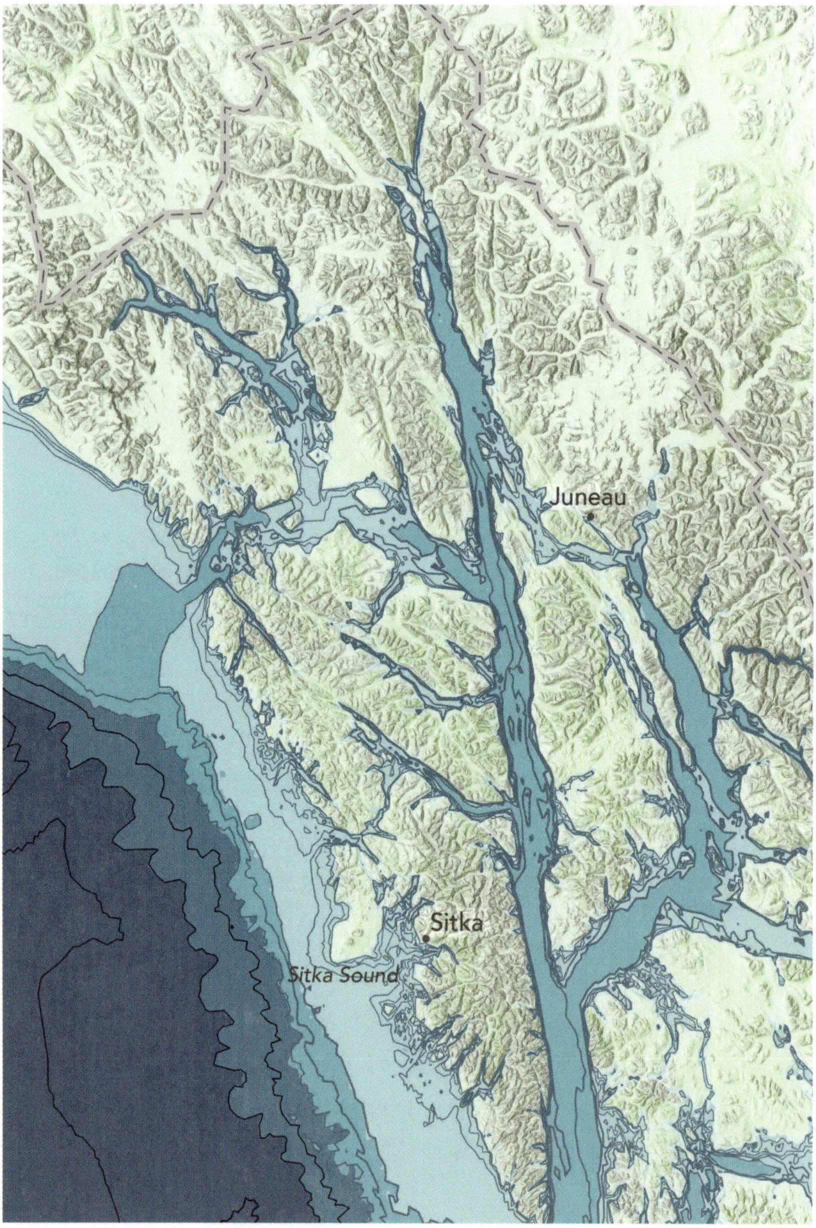
Juneau
Sitka
Sitka Sound

Depth Zone:

Trenches (>6,000 m)	Abyss (4,000 m–6,000 m)	Midnight Zone (2,000 m–4,000 m)	Midnight Zone (1,000 m–2,000 m)	Twilight Zone (200 m–1,000 m)	Sunlight Zone (<200 m)	Land

Southeast Alaska

> *Bull kelp loves places where the ocean runs like rivers. The intricacies of the SE geography creates so much diverse shoreline that bull kelp can find its happy place no matter how picky it is.*
>
> —Matt Kern, Barnacle Foods

Bull kelp is very happy in Southeast Alaska, the funny panhandle of the United States that sticks down into Canada. As in British Columbia, there are countless thin passageways of water between the thousands of islands that make up this unique stretch of coastline. As you move toward the poles, the difference between low tide and high tide becomes greater. Near Juneau there are twenty-five-foot tides; enormous amounts of water are pulled through each of these constricted passageways, first in one direction and then the other, in the same six-hour cycles as everywhere else. As the fast-growing bull kelp uses up the nearby ocean nutrients, a fresh supply is brought in on each tide, sustaining the kelp's remarkable year of development. Bull kelp beds here are abundant, healthy, and in many places expanding. The infinite types of shoreline enable bull kelp to find its preferred habitat: protected, or open to swell; near freshwater glacial output or a river estuary (both of which bring added nutrients into the nearshore environment); rocky bottom or more cobbled. Also, both top predators are present in the kelp forests: Sea otters are plentiful, and sea stars are recovering from the wasting syndrome. Their prey is varied, but the urchins in their diet are the large red urchins and the smaller green urchins.

Are they here?

Bull kelp: Yes, in abundance
Green urchin: Some
Northern sea otter: Yes, lots
Sunflower sea star: Yes. They seem to have recovered here from Sea Star Wasting Disease.
June 2025 average sea-surface temperature at Juneau: 9.5°C / 49°F

Kodiak and Alognak Islands

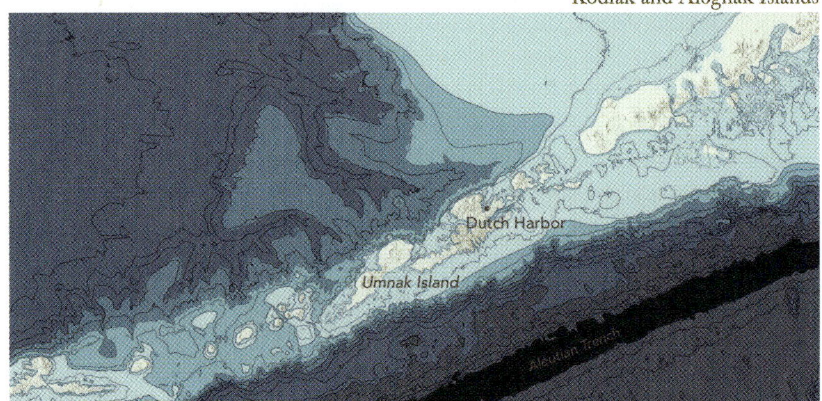

Eastern Aleutian Islands

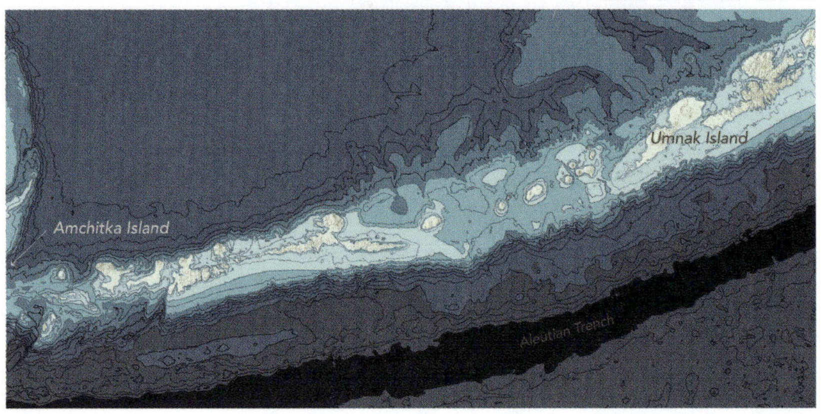

Western Aleutian Islands

Depth Zone:

Trenches (>6,000 m)

Abyss (4,000 m–6,000 m)

Midnight Zone (2,000 m–4,000 m)

Midnight Zone (1,000 m–2,000 m)

Twilight Zone (200 m–1,000 m)

Sunlight Zone (<200 m)

Land

Southwest Alaska

> *We were intrigued by the abrupt termination of* Nereocystis leutkeana *east of Samalga Pass and the exuberant expression of growth and reproduction in this population on the margin of this species' geographic range.*
>
> —Kathy Ann Miller, 1989

Nasquluq is the Alutiiq word for bull kelp. It comes from the word "head." The bull kelp in Alutiiq territory, from Prince William Sound into Cook Inlet and around the Kodiak Archipelago, is robust and healthy and has been used for centuries as a navigational aid indicating rocky outcrops and shallower waters; as fishing line made from the long singular stipe; and as fertilizer, creating soil for potato beds in the postcontact villages of Native and Russian-descended peoples. Around Kodiak Island, the northern sea otter population has held steady for decades. Bull kelp is abundant, with giant kelp becoming more apparent in recent years. A bit to the west, however, the sea otter population mysteriously crashed in the 1990s. Their numbers are not rising, and kelp habitat in the western Aleutians is impoverished and has commonly reverted to urchin barrens.

Bull kelp farming is being pioneered and refined on the southern coast of Alaska for commercial food and fertilizer products and to learn about bull kelp's mysterious developmental needs as an annual species. Bull kelp does not grow like sugar kelp, the most commonly farmed kelp in Alaska's growing kelp mariculture sector, so it requires specific experimental designs and research. It does not divulge its secrets easily.

Are they here?

Bull kelp: Yes, in abundance along the southern coast of Alaska, but in decline in the Aleutians

Green urchin: Some

Northern sea otter: Yes, lots

Sunflower sea star: Yes. They were not affected so badly by Sea Star Wasting Disease.

June 2025 average sea-surface temperature at Kodiak: 8.5°C / 47°F

Afterword

Forests on land have been studied intensively for hundreds, if not thousands, of years, whereas bull kelp forests of the North Pacific have been studied for only decades. It is hard to get down steep bluffs and into rough seas and cold oceans to do the studying. But the bull kelp forest is beginning to be recognized by the general public as foundational to the biodiversity of coastal ocean waters. As we humans try to shift from systems of resource extraction to the project of rebuilding resilience into natural systems, there is wonderful work at hand. Kelp forest restoration and urchin removal projects are becoming a major focus for agencies, academics, tribes, and artists alike. Research into kelp life cycles and genomics is finally getting the funding and attention it deserves. Partnerships and collaborations are offering inspiration and mentorship from one region to another. Bull kelp farming is being explored as a new industry for human foodways and product development. The 30x30 initiative (conserve 30 percent of terrestrial and marine habitat by 2030) is taking hold in regional, state, and national (we hope) programs that have real teeth for action. There is even a cohort of mermaids who don neoprene tails and seashell bikinis and become ambassadors for underwater worlds. Their call to action is "Help the kelp! Clear the urchins!" My partner, Marianna Leuschel, and I have felt attitudes shift since our ocean literacy campaign began in 2021 and our bull kelp web story was launched: Bull kelp conversation, concern, research, and cross-regional collaboration are bubbling with energy. Outreach to coastal communities through art exhibits, food extravaganzas, film festivals, and science fairs is now recognized as an essential component of kelp restoration.

But the underwater world of bull kelp still feels like a new realm to understand, requiring new collaborations, new networks of learning, and cooperation. Tribes, marine labs, academics, NGOs, community centers, artists, schoolteachers, town managers, and state and federal agencies are working to develop systems to help bull kelp recover. The young scientists impassioned by this work are also artists, and the young artists using the bull kelp forests as their muse, subject, and palette are also scientists. A spirit of cooperation abounds.

Despite the current defunding of science initiatives of all sorts, and the dishonoring of our natural world, I hope that this new vigor around the bull kelp forest will persist and that the organisms that depend on the kelp forest and that it depends on—the sea otters, sea stars, understory kelps and seaweeds, abalone, and sea urchins—might kindle our imaginations and charm us into changing how we relate to the natural world. We are a top predator alongside the sea otter and sunflower sea star, so we must act accordingly, asking ourselves at every turn, "Are we, as humans, inserting resilience *into* the kelp forest, or mining it out?" Beauty and artmaking can point us in the right direction.

Acknowledgments

This book is the result of an art and design journey that began in 2021 when I met Marianna Leuschel. We established Above/Below as an ocean artists' collective of sorts, and decided to transform my original bull kelp book-in-the-hand proposal into the web-based book titled *The Mysterious World of Bull Kelp* (bullkelp.info). We immediately asked Ellen Litwiller to join us, and she made the amazing home-page illustration and the character icons for the website. Marianna's vast experience running a communications design agency brought not only her expertise as a designer and editor to the project but also a wonderful team, including Civiane Chung, Alexandra Hammond, and Chad Upham, who helped us bring the website to life. We have just completed the final content for the website, including a section on kelp restoration and aquaculture. This book is a much-streamlined version of the web story, and we hope it will pique your curiosity to dive deeper into the bull kelp forest over on the digital version. Ellen has created new illustrations and revised others to work on the page. Marianna has been a first-line reader and editor of the material and a constant partner in getting this version over the finish line. Building collaborative capacity around ocean storytelling is what Above/Below is all about, and working as a team with these two has been a privilege and a whole lot of fun. On the About page of the website, Above/Below acknowledges the network of knowledge holders and the funders that made building the web story possible, but two names stand out. Sarah Allen, retired national parks biologist, and Rosa Laucci, marine program manager for the Tolowa Dee-Ni' Nation, were invaluable final editors.

In telling seaweed and kelp stories over many years, I have had the good fortune to work with and get to know a wonderful cadre of bull kelp scientists (phycologists), divers, researchers, cultivation experts, and photographers. I must always thank the Phycological Society of America, which has made a home for an artist in a scientific conference for so many years, and which has bestowed its Tiffany Award on my two seaweed books and now the bull kelp website. I am grateful for their generosity of spirit and for the many graduate students embracing the excitement of art and science working together to serve the seaweeds. I thank Tristin McHugh and the team at The Nature Conservancy that is working so hard to help the bull kelp on the Mendocino coast and for trusting Above/Below with the outreach and storytelling for that kelp restoration project. An ocean of gratitude goes to Heyday and especially Gayle Wattawa for taking a gamble on another seaweed book, which, despite the burgeoning interest in seaweed and kelp, still qualifies as niche.

And finally, I thank all the bull kelp enthusiasts whose love for this beautiful organism comes through in so many creative ways, proving that art and science together are a powerful way to connect, as humans, to the ocean world below the waves.

About the Contributors

Josie Iselin is an artist, author, and designer who has been telling seaweed and kelp stories for over a decade. Her two books *An Ocean Garden: The Secret Life of Seaweed* (2014, reprinted 2023) and *The Curious World of Seaweed* (Heyday, 2019) tap her profound understanding of seaweed natural history and her deep connections within the seaweed science community. Her work is an extraordinary expression of both art and science. Josie directs content development for the Above/Below campaign and is the lead author of the campaign's web story, *The Mysterious World of Bull Kelp*. She teaches in the School of Design at San Francisco State University.

Marianna Leuschel was the director of L Studio, a communications design agency, for twenty years and currently runs New Agency to develop communication strategies and design campaigns for a range of clients in ocean and land stewardship. She is particularly skilled at envisioning what hasn't yet been done and working with multiple stakeholders to create stories about our human relationship to the natural world and our responsibility as stewards. Marianna now leads the campaign for Above/Below, working with creative teams to generate new ideas for communication outreach, engagement strategies, and implementation of campaign initiatives.

Ellen Litwiller is a freelance illustrator whose work brings art and science together in imaginative ways. She loves exploring how creativity and curiosity intersect, using a variety of mediums to tell stories that are both visually striking and scientifically accurate. She began her career creating exhibits for natural history museums, where she worked as a muralist, illustrator, model maker, and preparator. With years of hands-on experience in exhibition design and installation, she developed a deep appreciation for detail and storytelling. Through collaboration with scientists, she enjoys the shared curiosity that unites art and science—both rooted in observation and appreciation of the world around us and the universe beyond.

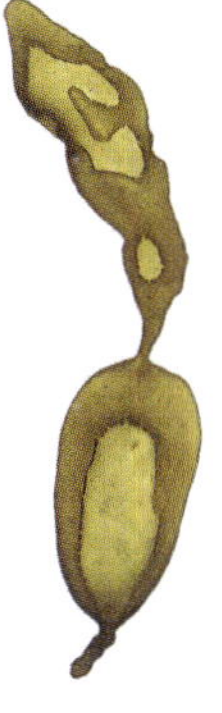

A Note on Type

This book is set in Caslon, the namesake typeface of the London-based typefounder and engraver William Caslon the Elder. First released in 1722, Caslon builds on the designs of seventeenth-century Dutch typefaces, and it remains popular today for its classic, organic style. The section heads are set in Minion Pro, and the secondary section heads are set in Avenir Next.

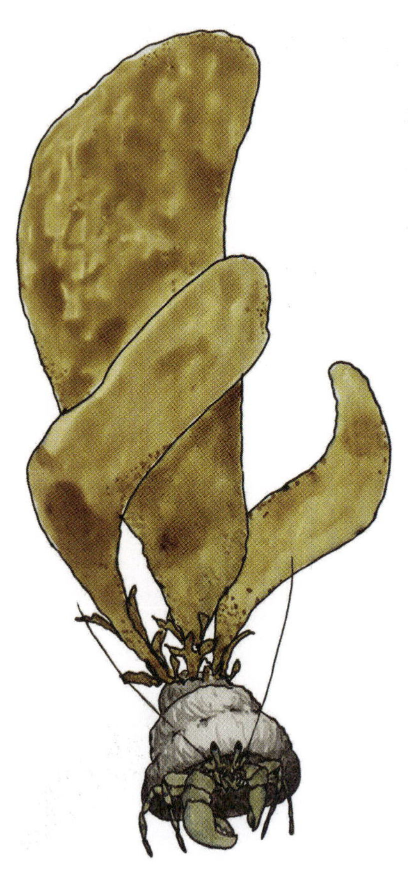